A contribution of the Ground-Water Resources Program

Ground-Water Availability in the United States

By Thomas E. Reilly, Kevin F. Dennehy, William M. Alley, and William L. Cunningham

Circular 1323

U.S. Department of the Interior
U.S. Geological Survey

U.S. Department of the Interior
DIRK KEMPTHORNE, Secretary

U.S. Geological Survey
Mark D. Myers, Director

U.S. Geological Survey, Reston, Virginia: 2008

For product and ordering information:
World Wide Web: *http://www.usgs.gov/pubprod*
Telephone: 1-888-ASK-USGS

For more information on the USGS--the Federal source for science about the Earth, its natural and living resources, natural hazards, and the environment:
World Wide Web: *http://www.usgs.gov*
Telephone: 1-888-ASK-USGS

Suggested citation:
Reilly, T.E., Dennehy, K.F., Alley, W.M., and Cunningham, W.L., 2008, Ground-Water Availability in the United States: U.S. Geological Survey Circular 1323, 70 p., also available online at *http://pubs.usgs.gov/circ/1323/*

Library of Congress Cataloging-in-Publication Data

Ground-water availability in the United States / by Thomas E. Reilly ... [et al.].
p. cm. -- (U.S. Geological Survey circular ; 1323)
Includes bibliographical references and index.
ISBN 978-1-4113-2183-0 (alk. paper)
1. Groundwater--United States. I. Reilly, Thomas E. II. Title: Groundwater availability in the United States.
GB1015.G758 2008
553.7'90973--dc22

2008026582

Foreword

Ground water is among the Nation's most important natural resources. It provides half our drinking water and is essential to the vitality of agriculture and industry, as well as to the health of rivers, wetlands, and estuaries throughout the country. Large-scale development of ground-water resources with accompanying declines in ground-water levels and other effects of pumping has led to concerns about the future availability of ground water to meet domestic, agricultural, industrial, and environmental needs. The challenges in determining ground-water availability are many. This report examines what is known about the Nation's ground-water availability and outlines a program of study by the U.S. Geological Survey Ground-Water Resources Program to improve our understanding of ground-water availability in major aquifers across the Nation. The approach is designed to provide useful regional information for State and local agencies who manage ground-water resources, while providing the building blocks for a national assessment. The report is written for a wide audience interested or involved in the management, protection, and sustainable use of the Nation's water resources.

Robert M. Hirsch, Associate Director for Water
U.S. Geological Survey

Blue Springs, Blue Springs State Park, Barbour County, Alabama
Photograph by Alan M. Cressler, USGS

Contents

Boxes

Artesian well at Woonsocket, well throws a 3-inch-wide stream to height of 97 feet, Sanborn County, South Dakota, Circa 1900 (Darton, N.H., 1905, Plate 60-B).

Ground-Water Availability in the United States

By Thomas E. Reilly, Kevin F. Dennehy, William M. Alley, and William L. Cunningham

Introduction

Ground water is one of the Nation's most valuable natural resources. It occurs almost everywhere beneath the Earth's surface and is a major source of water supply worldwide. Ground water has a crucial role in sustaining streamflow between precipitation events and especially during protracted dry periods. In addition to human uses, many ecosystems are dependent on ground-water discharge to streams, lakes, and wetlands.

Although humans have been digging wells and tunnels for water supply for thousands of years, extensive use of ground water is relatively recent, with the advent of rural electrification and more effective drilling and pumping technologies during the past 75 years. A growing awareness of ground water as a critical natural resource leads to some basic questions. How much ground water do we have? Are we running out? Where are ground-water resources most stressed by human development? Where are the resources most available for future supplies? Although these questions seem simple, providing the answers is complex because a meaningful assessment of ground-water availability in the United States requires a multidisciplinary evaluation of the hydrologic system, as well as an understanding of the different water issues that exist across the Nation. Furthermore, the information available to support a broad assessment of the resource varies across the Nation (Alley, 2006).

O.E. Meinzer in 1934.

During the past century, several ground-water assessments have been completed by the U.S. Geological Survey (USGS) on a national scale. The first of these assessments was completed by O.E. Meinzer (1923) who has been called the "father of ground-water hydrology" (Lohman, 1986, p. 51). Meinzer's publication was followed several decades later with State-by-State summaries on ground-water resources (McGuinness, 1951 and 1963); by summary appraisals for 21 regions of the Nation in the 1970s (U.S. Geological Survey Professional Papers 813A–U); a State-by-State summary (U.S. Geological Survey, 1985); and by the Regional Aquifer-System Analysis (RASA) Program in which 25 of the Nation's most important regional ground-water systems were evaluated (Sun and Johnston, 1994). These national and regional evaluations have improved our knowledge about the Nation's ground-water resources. Repeated evaluations of the resource through time are needed as new information on ground-water resources and connected surface-water systems becomes available; new methods and technologies for resource assessment are developed; and the places ground water is used, water demands, and the issues of concern change with time.

> *"Finding scientific and technical solutions to problems of water availability and quality will require extensive cooperation and collaboration among Federal, State, and local agencies, private sector water experts, stakeholders, and the public..."*
>
> From "A Strategy for Federal Science and Technology to Support Water Availability and Quality in the United States," National Science and Technology Council (2007)

Environmental decision making has grown more complex with society demanding ample water for human use along with environmental protection and preservation at the same time. When O.E. Meinzer (1923) published his first national ground-water assessment, indoor plumbing still was not commonly used, low-capacity wells were sufficient for most purposes, and the population of the country was more dispersed. Today, lifestyles generally require large amounts of water and a complex infrastructure to deliver water to urban and suburban population centers. Even if water resources are abundant regionally, heavy water use in centralized areas can create local stresses. As water-related problems evolve in complex ways, an up-to-date and comprehensive evaluation of ground-water resources that builds on the foundation of previous studies is needed to meet society's ever-changing water demands.

A goal of ground-water resource assessment is to provide information on the current status of the resource that provides insights about the future availability of ground water. Ground-water management decisions in the United States are made at a local level, such as the State, municipality, or a special district formed for water-resources management. Many aquifer systems cross these political boundaries. Thus, a key role of national and regional assessments is to provide consistent and integrated information across political boundaries that is useful to those who use and manage the resource. The State and local agencies manage the water-resources system and collect and analyze local data. Federal scientific agencies support this function by developing methods of analysis and analyzing the water-resources system across political boundaries. This partnership between State and local agencies and the USGS enables the resource to be understood on a multi-State, regional, and national basis.

With these considerations in mind, the purpose of this report is to identify the challenges in determining ground-water availability, summarize the current state of knowledge from a national perspective, and outline an approach for developing the needed understanding of future water availability. This report is an outgrowth of a pilot study, National Assessment of Water Availability and Use, that began in 2005 at the request of Congress (Barlow and others, 2002). The report also builds on regional ground-water availability studies recently undertaken as part of the USGS Ground-Water Resources Program (Dennehy, 2005). The approach to national ground-water assessment described in the section "Regional-Scale Approach to National Assessment" of this report, is a key element of the water census of the United States, which has been proposed as a strategic science direction of the USGS (U.S. Geological Survey, 2007), as well as part of the proposed Federal science strategy to meet nationwide water challenges by the National Science and Technology Council (2007) Subcommittee on Water Availability and Quality.

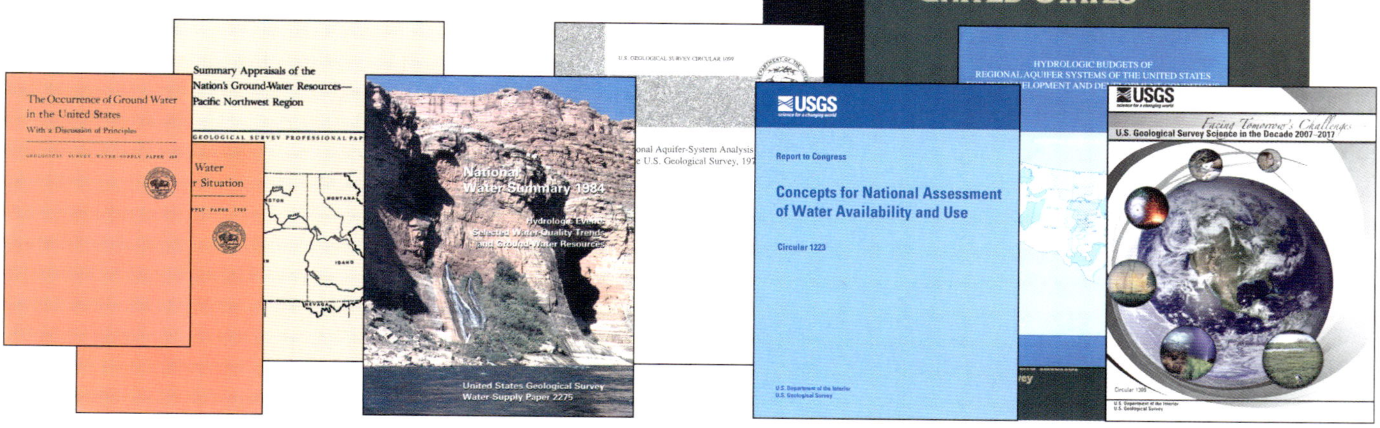

Challenges in Determining Ground-Water Availability

Although determining the amount of ground water available in the Nation may seem straightforward, it is actually quite complex. Some key difficulties are as follows:

1. In contrast to rivers and lakes, ground-water systems are hidden from direct observation and measurement,
2. The sources of water to ground-water systems and the time required for the effects of withdrawals to propagate through the system and be observed are different for each system,
3. The amount of detail (spatial scale) needed to describe the resource depends on the objectives and purpose of the desired information,
4. The amount of change in ground-water levels that is important is different for different ground-water systems,
5. Not all water pumped is consumed and much of the water pumped is redistributed and changes the ground-water flow system, and
6. The chemical quality of the water is important in determining its suitability (and thus its availability) for various uses.

These challenges are considered throughout this report.

Determining ground-water availability means more than calculating the volume of ground water underlying a particular area or within an aquifer. One must not only consider that some of the water may not be economically recoverable or of poor quality but also that ground water is connected to the rest of the hydrologic system. Ground-water withdrawals can and usually do affect the amount (and quality) of surface water. For example, depletion of a small part of the total volume of ground water in storage (sometimes only a few percent) can have substantial and undesirable effects on the availability of surface water that becomes the limiting factor to development of the ground-water resource (Alley, 2007). Increasingly, contributions of ground water to surface water are considered an important part of ecosystem needs, and in some cases, plant and animal communities depend partly or completely on ground water to maintain their current composition and function. Thus, an assessment of ground-water availability requires consideration of the response of the entire hydrologic system to ground-water withdrawals.

In discussions of ground-water availability, terms such as *ground-water mining* and *safe yield* frequently are used to indicate the status of the resource. The use of different terms can sometimes exacerbate the difficulties in answering questions about water availability because these terms can have different meanings. Four of these terms are defined in Box A.

As a foundation for the discussion in this report, it is helpful to consider the meaning of the terms *water availability* and *ground-water availability*. Although the quantities of water in a hydrologic system usually can be measured, computed, or estimated, water availability is a more elusive and multifaceted concept. Water availability is a function not only of the quantity and quality of water in a basin or aquifer system but also the physical structures, laws, regulations, and socioeconomic factors that control its demand and use. This report discusses physical and chemical characteristics that are important as indicators of ground-water availability. At the local level, these characteristics must be considered jointly with societal factors as determinants of actual ground-water availability and society's tolerance of the consequences of its use. Societal perspectives and constraints change with time just as the ground-water resource does (Alley and Leake, 2004).

A

Terms Used in Describing Ground-Water Availability

A number of terms are used to describe ground-water availability. For example, ground-water resources often are discussed in terms of their sustainability. As defined in Alley and others (1999), ***ground-water sustainability*** is the "development and use of ground water in a manner that can be maintained for an indefinite time without causing unacceptable environmental, economic, or social consequences." The definition of "unacceptable consequences" is largely subjective and may involve a large number of criteria. Furthermore, ground-water sustainability must be defined within the context of the complete hydrologic system of which ground water is a part. For example, what may be established as an acceptable rate of ground-water withdrawal with respect to changes in ground-water levels may reduce the availability of surface water to an unacceptable level. Determination of ground-water sustainability essentially is the end result of the public acceptance of the tradeoffs of development. A key role of hydrologists is to provide information, such as is described in this report, on the long-term consequences of pumping and other factors affecting ground-water resources that can aid societal decisions related to sustainability. We use the term ground-water sustainability in this report in this general context.

The term ***safe yield*** commonly is used in efforts to quantify sustainable ground-water development. The term should be used with respect to specific effects of pumping, such as water-level declines, reduced streamflow, and degradation of water quality. Alley and Leake (2004) describe the history of the term and its relation to the concept of sustainability. The term ***ground-water mining*** typically refers to a prolonged and progressive decrease in the amount of water stored in a ground-water system, as may occur, for example, in heavily pumped aquifers in arid and semiarid regions. Ground-water mining is a hydrologic term without connotations about water-management practices (U.S. Water Resources Council, 1980). The term ***overdraft*** refers to withdrawals of ground water from an aquifer at rates considered to be excessive and, therefore, carries the value judgment of overdevelopment. Thus, overdraft may refer to ground-water mining that is considered excessive as well as to other undesirable effects of ground-water withdrawals. The terms safe yield, ground-water mining, and overdraft are not used in this report.

When discussing water availability, the topic of ***well yields*** (that is, how much water a particular well can produce) frequently arises. Well yields, however, only address the efficiency with which a well will allow water to be removed from the ground-water system. The well must be placed in the context of the flow system around the well and the amounts of water being withdrawn by other wells in the area. Just because a well initially can pump a certain amount of water does not mean that the ground-water system can supply that amount of water indefinitely. An analysis of the ground-water system is required to determine the source of the water that is being withdrawn from the ground-water system and to determine if there is a sufficient amount of water available from these sources. Thus, well yields, in and of themselves, do not address regional ground-water availability.

Ground Water—The Hidden Resource

Ground water is a hidden resource, in that we cannot visually observe its movement and status. We must, therefore, gain information by measurements obtained from wells, by measurements of flow and water levels at its boundaries, and through indirect methods of measurement (such as surface geophysics) to assess the resource. These data are used to infer the actual occurrence and movement of ground water and to develop a conceptual model of the ground-water system. This conceptual model can be used to explain the extent of the ground-water system, the sources of water to the system, and the movement (rate and direction) of water through the inferred hydrogeologic units. The conceptual model can never be exact and is subject to uncertainty and error because of the indirect nature of the measurement methods and the complexities of the subsurface and natural systems in general.

Ground-water systems store and transmit water. One of the advantages of ground water is that it exists almost everywhere across the Nation and, thus, is available away from surface sources of water. This advantage enables communities, individual well owners, and irrigators the opportunity to obtain water without investing in pipelines and storage facilities. The water in a ground-water system is stored naturally in the pore space or fractures of the earth. As ground water is withdrawn at a well, the connected pore spaces or fractures serve as the pipeline to move the water from one part of the hydrogeologic system to where it is being withdrawn. Understanding the movement of water through the ground-water system and understanding the limits of the sources of water are key aspects of a ground-water availability assessment.

The amount of detail needed to describe and assess the resource depends on the objectives and purpose of the estimate. The pumpage of fresh ground water in the United States in 2000 was estimated to be approximately 83 billion gallons per day (Hutson and others, 2004), which is about 8 percent of the estimated 1 trillion gallons per day of natural recharge to the Nation's ground-water systems (Nace, 1960). From an overall national perspective, therefore, the ground-water resource appears ample. Throughout the Nation, however, the availability of ground water varies widely.

Ground-water data collection. Photographs by USGS staff: Alan M. Cressler, William L. Cunningham, Kevin F. Dennehy, Matthew J. Gilbert, Edward H. Martin, Lester J. Williams, and Douglas D. Zettwoch.

Time Scales of Ground-Water Systems

There are two main processes to consider in determining time scales of ground-water systems. One is the time it takes water levels to respond to changes in stress (such as pumping) on the ground-water system; the other is the time it takes the water to travel through the ground-water system. The time frame of changes in water levels depends on how quickly the change in water levels propagates through the system after water is removed from storage. The time of travel of water flowing through the system depends on the velocity of the water and the distance between the recharge and discharge boundaries. These two times are very different for most ground-water systems.

The amount of water coming from different sources to a discharging well changes through time until, if possible, a new steady-state or equilibrium condition is established. For example, figure 1 illustrates the sources of water in a simple idealized stream-aquifer system supplying one well. At the start of pumping, 100 percent of the water supplied to the well comes from ground-water storage. Over time, the dominant source of water to the well changes from ground-water storage to surface water. The surface-water source for purposes of discussion here is a stream, but it may be another surface-water body, such as a lake or wetland. The source of water to a well from a stream can be either decreased ground-water discharge to the stream or increased flow (recharge) from the stream into the ground-water system. The streamflow reduction in either case is referred to as streamflow capture. The adjustments to pumping of an actual hydrologic system may take place over many years, decades, or longer, depending on the physical characteristics of the aquifer, degree of connection between the stream and aquifer, and locations and pumping history of wells.

Most ground-water systems are much more complex than implied in figure 1; for example, the system may comprise many wells pumping from an aquifer at varying pumping rates and at different locations within the ground-water flow system. From an availability perspective, the key point is stated by Theis (1940) as, "All water discharged by wells is balanced by a loss of water somewhere." The ground-water system is part of the encompassing hydrologic cycle, and water taken from the ground-water system has to come from storage or other parts of the hydrologic cycle that are connected to the ground-water system. These connected systems might be surface-water bodies, such as streams, or they could be plant communities that use the water for growth, such as riparian vegetation. Pumping decisions made today will affect water availability for the surrounding ecosystem; however, these effects may not be fully realized for many years.

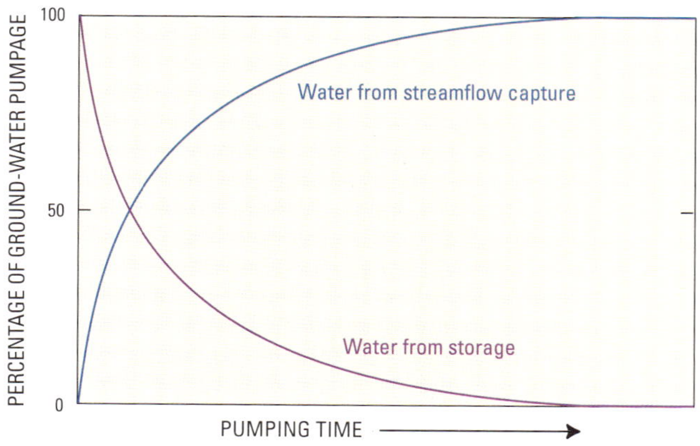

Figure 1. The principal source of water to a well can change with time from ground-water storage to capture of streamflow (modified from Alley and others, 1999).

The time of travel of the water flowing through a ground-water system is different from the time it takes the water levels to respond and is widely variable, ranging from less than a day to more than a million years (Bentley and others, 1986). As a result, water stored within the system can range in age (the time since recharge) from recent precipitation to water trapped in the sediments as they were deposited in geologic time. For the 48 contiguous States, Nace (1960) estimated that there are about 60,000 trillion gallons of ground water in storage. Assuming net recharge of 1 trillion gallons a day, about 160 years of recharge is stored in the ground. In the humid East, the average storage time of ground water is probably shorter, and in the arid and semiarid central and western States, it is probably longer.

For large systems with long flow paths and large travel times, past climate variability may be important in the initial saturation of the aquifer and the distribution of water in storage. Future climate variability and change also may be important in determining water availability. For example, in the mountains of the western United States, precipitation in recent decades has come more frequently in the form of rain rather than snow (Knowles and others, 2006), and snowpacks have thinned (Mote and others, 2005), which likely changes the quantity and distribution of recharge in the mountains as well as recharge from mountain runoff. As climate changes, ground-water systems will respond, but the effects may take long periods of time to fully develop.

What Do We Know About Ground-Water Availability in the United States?

In order to determine the availability of water, we need information about the resource (the supply) and about its use (the demand). The amount of detail needed is dependent on the objectives and scale of the analysis. On a national scale, we know quite a bit about the Nation's ground-water resource; however, much of the information is generalized and has limitations when attempts are made to plan for the future. Even though national-scale information has its limitations, it provides a framework for a systematic comparison of the resource across the Nation. This section of the report reviews the information available on a national scale and also indicates the limitations of that information when determining water availability.

Location and Description of Major Aquifers

An aquifer is a geologic formation, a group of formations, or a part of a formation that contains sufficient saturated permeable material to yield significant quantities of water to wells and springs. The areal and vertical location of the major aquifers is fundamental to the determination of ground-water availability for the Nation.

The location, hydrologic characteristics, and geologic characteristics of the principal aquifers throughout the 50 States, Puerto Rico, and the U.S. Virgin Islands are described in the Ground Water Atlas of the United States (Miller, 2000; *http://capp.water.usgs.gov/gwa/*). The ground-water information summarized in the atlas has been collected over many years by the USGS and other partner agencies. The Atlas provides key descriptive information in a regional and national context.

A two-dimensional map representation of the principal aquifers of the Nation (U.S. Geological Survey, 2003) is shown in figure 2. The map, which is derived from the Ground Water Atlas of the United States, indicates the areal extent of the uppermost principal aquifers on a national scale. Although the map is two dimensional, it provides a useful visual representation of the Nation's complex three-dimensional ground-water resource. In some places, other productive aquifers underlie those shown on the map. For example, the highly productive limestone that forms the Floridan aquifer system of the southeastern United States underlies the entire Florida Peninsula and extends into Georgia, Alabama, and South Carolina. Only small areas of this aquifer system are shown on the map, because it is covered in many places by younger sand aquifers. Likewise, some aquifers in sedimentary rocks are overlain by confining units and extend into the subsurface beyond the areas shown on the map. Local aquifers, such as stream-valley aquifers that might overlie the aquifers mapped in figure 2, are not shown because of the scale of the figure.

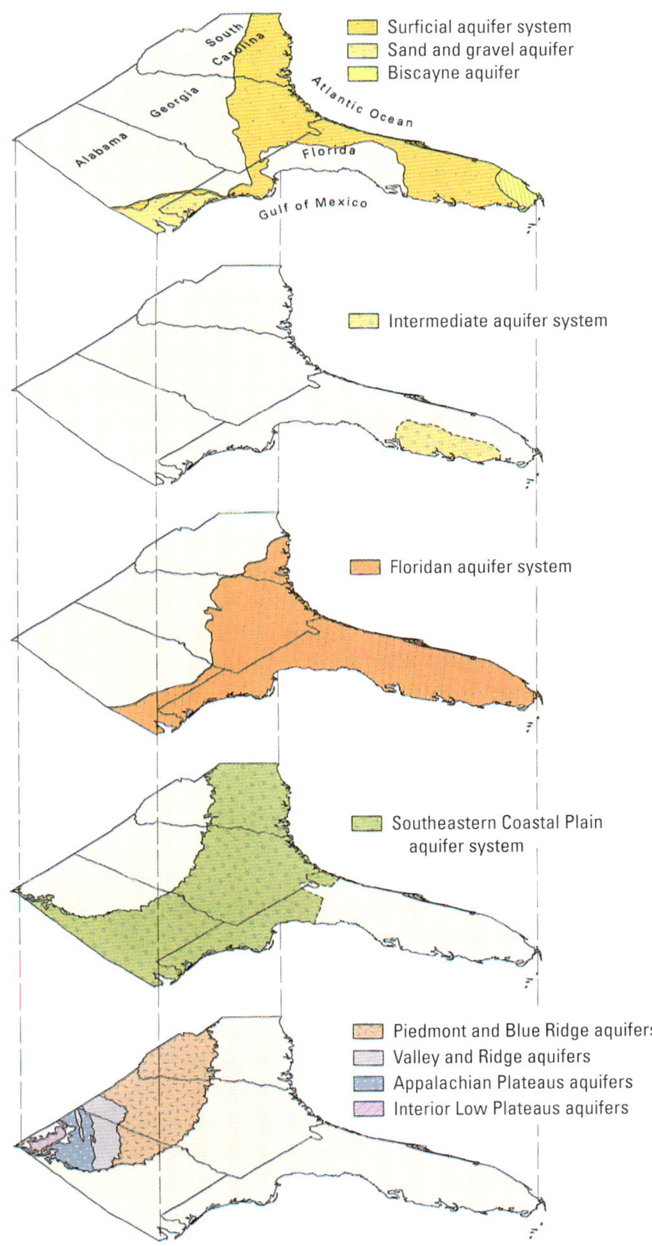

Perspective view of overlapping aquifer systems in the Alabama, Florida, Georgia, and South Carolina area as described in text and shown in figure 2 (from Miller, 2000).

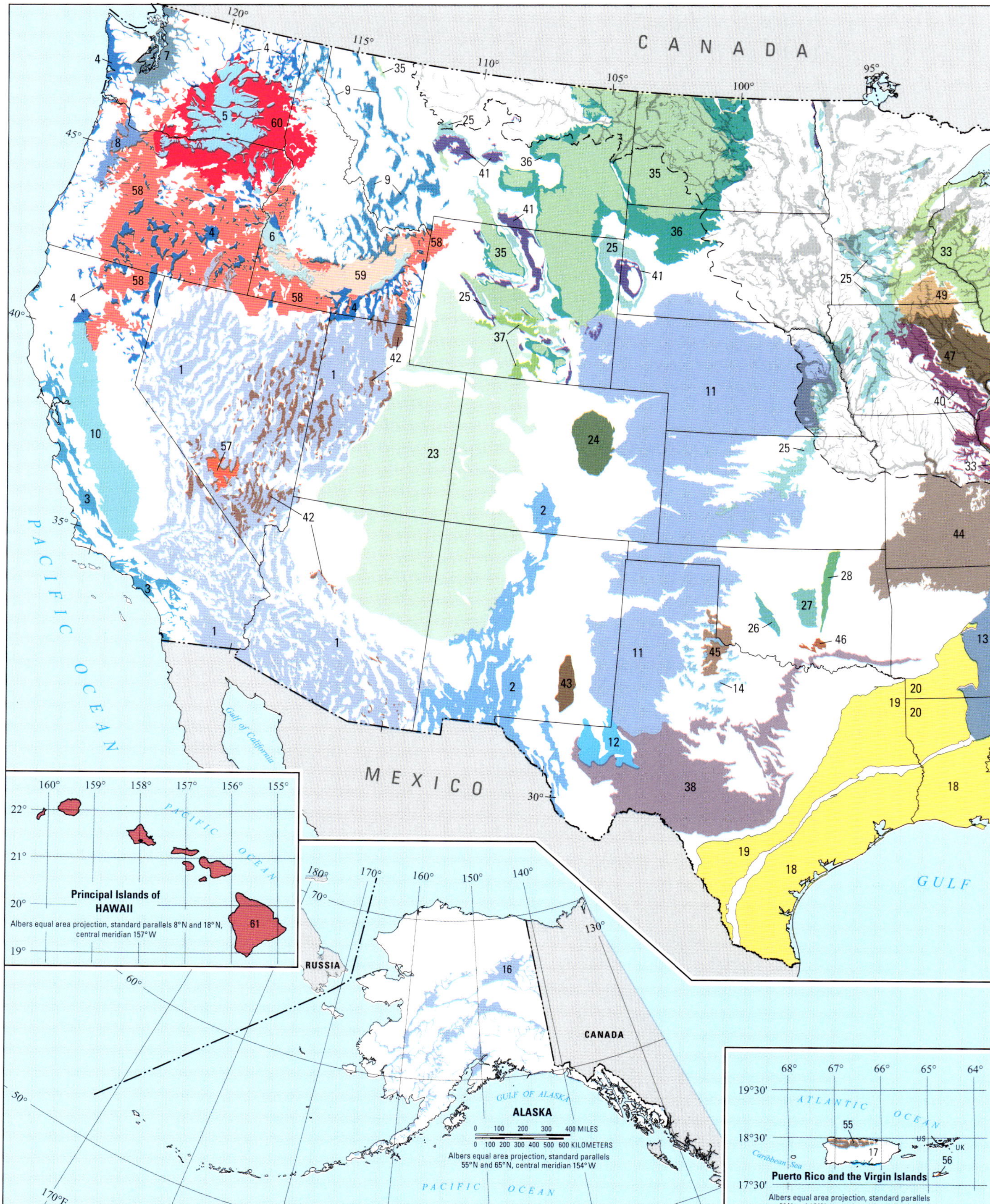

Figure 2. Principal aquifers of the United States (modified from Principal Aquifers, U.S. Geological Survey, 2003).

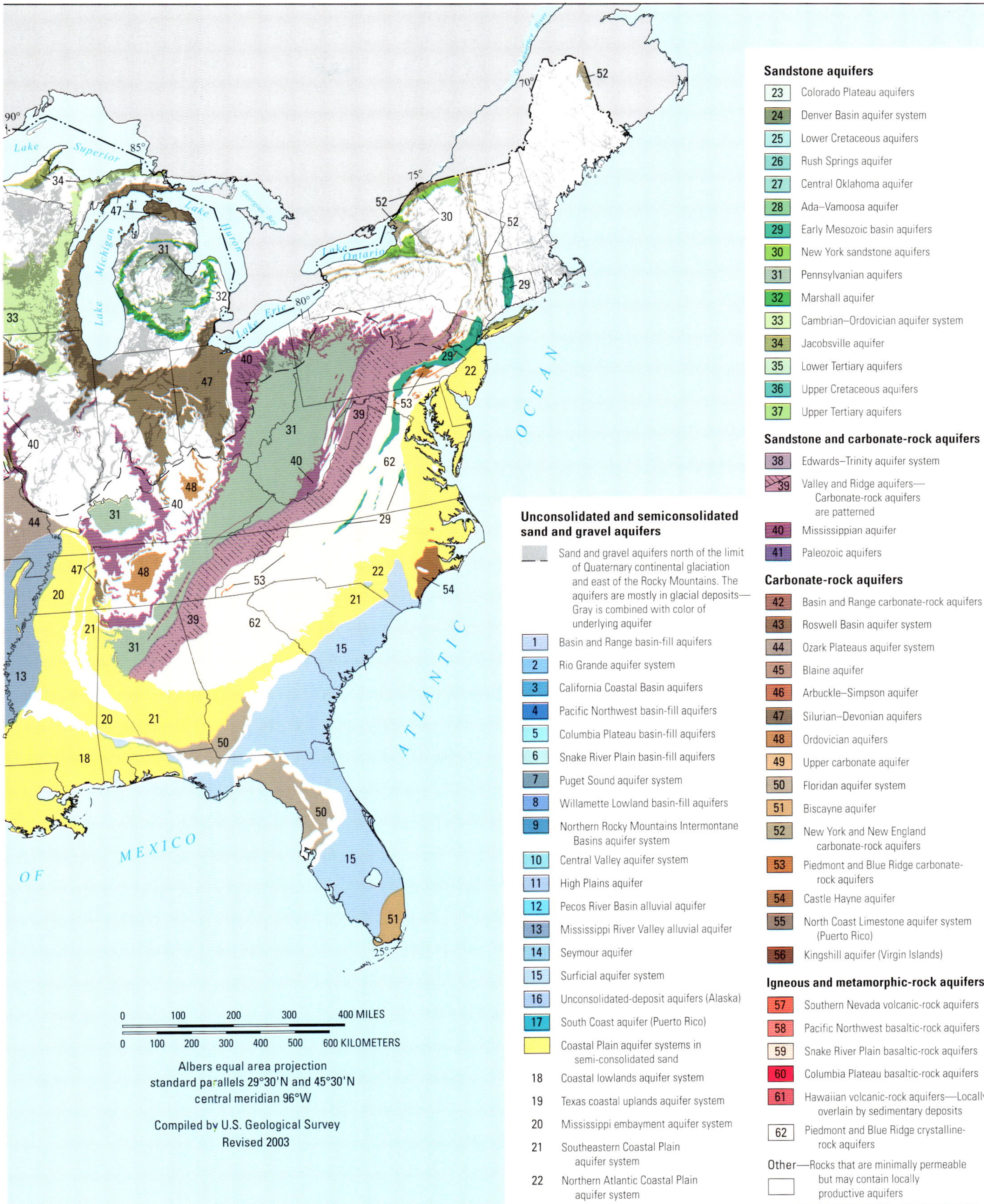

Lake Superior
Lake Michigan
Lake Huron
Georgian Bay
Lake Ontario
Lake Erie
St. Lawrence River
ATLANTIC OCEAN
OF MEXICO
Unconsolidated and semiconsolidated sand and gravel aquifers
Sand and gravel aquifers north of the limit of Quaternary continental glaciation and east of the Rocky Mountains. The aquifers are mostly in glacial deposits—Gray is combined with color of underlying aquifer
1 Basin and Range basin-fill aquifers
2 Rio Grande aquifer system
3 California Coastal Basin aquifers
4 Pacific Northwest basin-fill aquifers
5 Columbia Plateau basin-fill aquifers
6 Snake River Plain basin-fill aquifers
7 Puget Sound aquifer system
8 Willamette Lowland basin-fill aquifers
9 Northern Rocky Mountains Intermontane Basins aquifer system
10 Central Valley aquifer system
11 High Plains aquifer
12 Pecos River Basin alluvial aquifer
13 Mississippi River Valley alluvial aquifer
14 Seymour aquifer
15 Surficial aquifer system
16 Unconsolidated-deposit aquifers (Alaska)
17 South Coast aquifer (Puerto Rico)
Coastal Plain aquifer systems in semi-consolidated sand
18 Coastal lowlands aquifer system
19 Texas coastal uplands aquifer system
20 Mississippi embayment aquifer system
21 Southeastern Coastal Plain aquifer system
22 Northern Atlantic Coastal Plain aquifer system
Sandstone aquifers
23 Colorado Plateau aquifers
24 Denver Basin aquifer system
25 Lower Cretaceous aquifers
26 Rush Springs aquifer
27 Central Oklahoma aquifer
28 Ada–Vamoosa aquifer
29 Early Mesozoic basin aquifers
30 New York sandstone aquifers
31 Pennsylvanian aquifers
32 Marshall aquifer
33 Cambrian–Ordovician aquifer system
34 Jacobsville aquifer
35 Lower Tertiary aquifers
36 Upper Cretaceous aquifers
37 Upper Tertiary aquifers
Sandstone and carbonate-rock aquifers
38 Edwards–Trinity aquifer system
39 Valley and Ridge aquifers—Carbonate-rock aquifers are patterned
40 Mississippian aquifer
41 Paleozoic aquifers
Carbonate-rock aquifers
42 Basin and Range carbonate-rock aquifers
43 Roswell Basin aquifer system
44 Ozark Plateaus aquifer system
45 Blaine aquifer
46 Arbuckle–Simpson aquifer
47 Silurian–Devonian aquifers
48 Ordovician aquifers
49 Upper carbonate aquifer
50 Floridan aquifer system
51 Biscayne aquifer
52 New York and New England carbonate-rock aquifers
53 Piedmont and Blue Ridge carbonate-rock aquifers
54 Castle Hayne aquifer
55 North Coast Limestone aquifer system (Puerto Rico)
56 Kingshill aquifer (Virgin Islands)
Igneous and metamorphic-rock aquifers
57 Southern Nevada volcanic-rock aquifers
58 Pacific Northwest basaltic-rock aquifers
59 Snake River Plain basaltic-rock aquifers
60 Columbia Plateau basaltic-rock aquifers
61 Hawaiian volcanic-rock aquifers—Locally overlain by sedimentary deposits
62 Piedmont and Blue Ridge crystalline-rock aquifers
Other—Rocks that are minimally permeable but may contain locally productive aquifers
0 100 200 300 400 MILES
0 100 200 300 400 500 600 KILOMETERS
Albers equal area projection
standard parallels 29°30'N and 45°30'N
central meridian 96°W
Compiled by U.S. Geological Survey
Revised 2003

Some of the principal aquifers shown in figure 2 are systems of multiple aquifers. For example, the "Northern Atlantic Coastal Plain aquifer system" is identified on the national map as a principal aquifer composed of semiconsolidated sand that is present in several States, including southern New Jersey. At a regional level, however, the "Northern Atlantic Coastal Plain aquifer system" is actually a system of aquifers and confining units (Voronin, 2004) as shown in figure 3 for the New Jersey Coastal Plain. At a local level, the distribution of sands and clays are even more variable than those shown at the regional level.

Thus, we have a broad understanding of where the principal water-bearing formations in the United States are located. The level of detailed understanding of ground-water systems varies widely across the United States. The principal aquifers, shown in figure 2, provide a framework to classify and study ground-water systems regionally. These potential areas for regional investigations can be prioritized in conjunction with other information on sources and uses of water, as discussed later in this report.

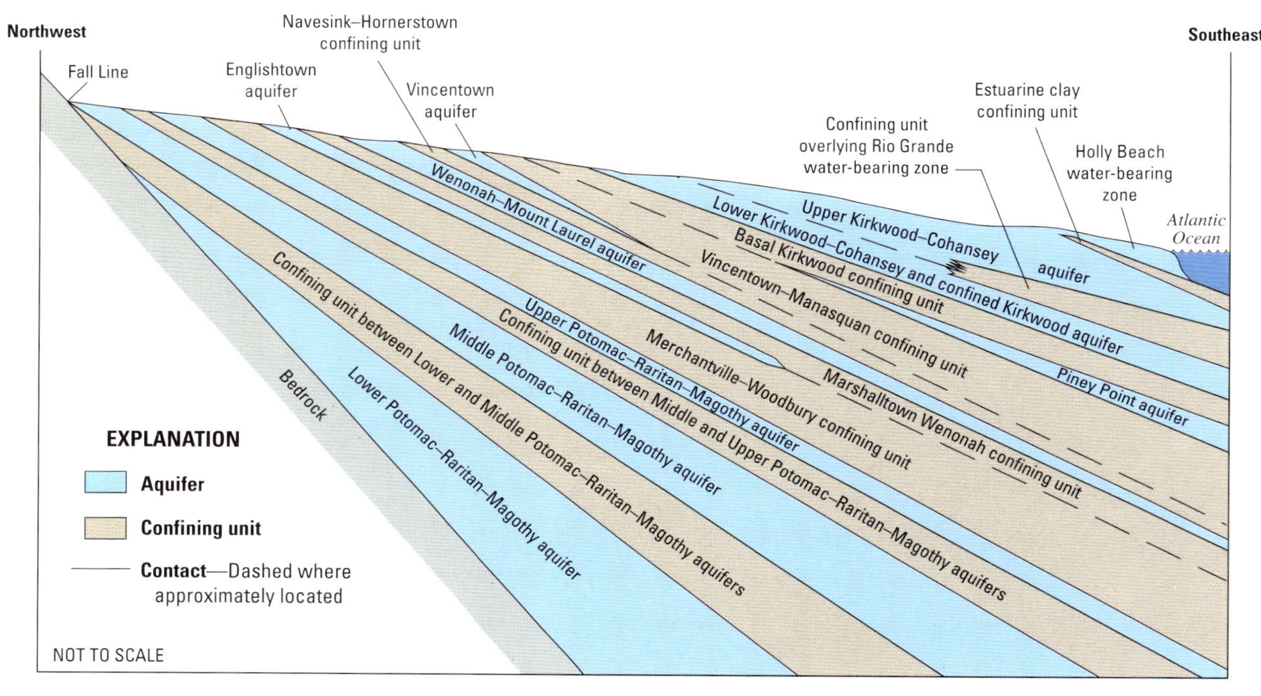

Figure 3. Generalized hydrogeologic section of the New Jersey Coastal Plain (modified from Voronin, 2004).

Water Use

The USGS has partnered with State and local agencies to compile estimates of ground-water and surface-water withdrawals for the Nation at 5-year intervals since 1950. The data currently are compiled at the county, State, and national levels for eight categories of water use—public supply, domestic, irrigation, livestock, aquaculture, self-supplied industrial, mining, and thermoelectric power. The most recent compilation is for the year 2000 (Hutson and others, 2004).

> *"The United States should accurately assess the quantity and quality of its water resources, should accurately measure how water is used, and should know how water supply and use change over time."*
>
> From "A Strategy for Federal Science and Technology to Support Water Availability and Quality in the United States," National Science and Technology Council (2007)

The precision of water-use data over time is affected by the status of State water-use reporting programs and the development of techniques for estimating water use. Programs to collect water-use data in each State are highly variable and are summarized in a review of the USGS water-use program (National Research Council, 2002).

Some water-use data, such as public supply for household uses and withdrawals by some industrial users, are obtained by direct measurement, and some are estimated as the amount reported or allowed by permit. Many uses, such as for self-supplied domestic use, irrigation, and some industries, commonly are estimated using coefficients that relate water use to another characteristic. For example, water use for a particular type of industry might be estimated by using information on employment or production in terms of gallons per day per employee or per unit of product.

Despite the inherent differences in data-collection methods from year to year and State to State, the USGS water-use data represent a unique record of withdrawals over time and throughout the country. The data provide a broad-based 50-year history of changes in water withdrawals. For example, the data indicate that ground-water withdrawals more than doubled between 1950 and 1975 but subsequently have remained fairly steady, that the percentage of ground water compared to surface water for public supply increased from about 26 percent in 1950 to about 40 percent in 1985 and has remained just under 40 percent during the past two decades, and that the percentage of ground water compared to surface water for irrigation increased from about 23 percent in 1950 to about 42 percent in 2000 (Hutson and others, 2004). In the most recent compilation (for year 2000), ground-water withdrawals for irrigation accounted for about two-thirds of total ground-water withdrawals (Hutson and others, 2004). The temporal trends of ground-water withdrawals for irrigation have been somewhat different between the western and eastern States (fig. 4).

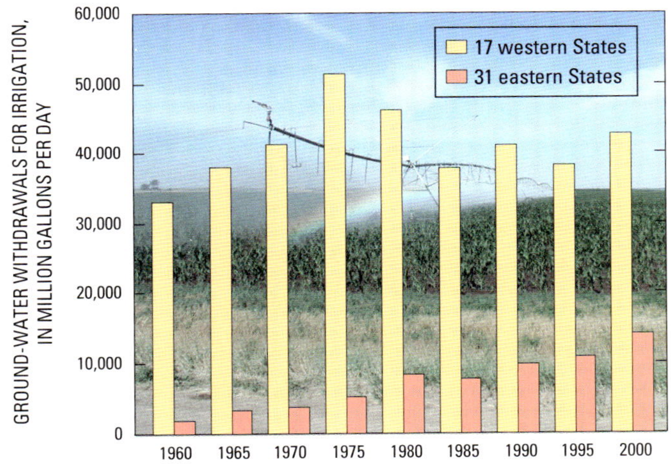

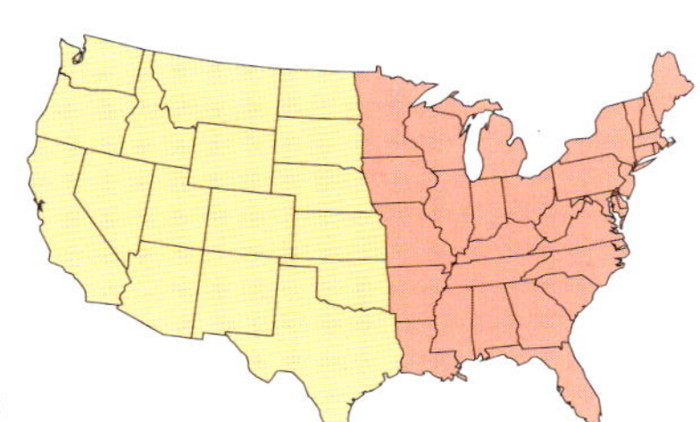

Figure 4. Ground-water withdrawals for irrigation in the western and eastern conterminous United States. Ground-water withdrawals for irrigation decreased in the western States in recent decades as a result of expanding urban areas, an increase in dryland farming, and increased efficiencies of application. In contrast, ground-water withdrawals for irrigation in the eastern half of the country increased steadily over the same period, in part, as a supplemental source of water to protect against dry periods. (Data compiled from U.S. Geological Survey Circulars titled "Estimated use of water in the United States," published in 5-year intervals for the years 1960 to 2000.)

As part of the 2000 compilation of water-use data, estimates of ground-water withdrawals were made for 66 principal aquifers in the United States for three major categories of water use—public supply, irrigation, and self-supplied industrial (Maupin and Barber, 2005). The results indicate that 20 principal aquifers (including an "other" principal aquifer category) account for about 90 percent of the ground-water withdrawals in the United States for the three major categories combined (fig. 5).

In estimating ground-water use, it is important to recognize that not all the water pumped is consumed. When water is pumped from the ground and used, the water molecules are not destroyed; the water is simply moved to different places. For example, when water is used for self-supplied domestic use, some of it is consumed and some of it is redistributed back into the environment. Solley and others (1998, p. 24) estimated that only 26 percent of the water used for self-supplied domestic use is consumed. Consumed water is assumed to be evaporated, transpired, incorporated into products or crops, consumed by humans or livestock, or otherwise removed from the immediate water environment. The rest of the water gets redistributed back into the environment, such as sewage disposal into streams and additional recharge from excess irrigation. Even the water consumed, however, is not really lost; it goes into the atmosphere or into products or living tissue. When analyzing the amount of ground-water available, it is important to consider where the water pumped will end up. Thus, ideally, information on ground-water use should include estimates of consumptive use and return flow as well as withdrawals, but this type of information can be difficult to estimate for many uses.

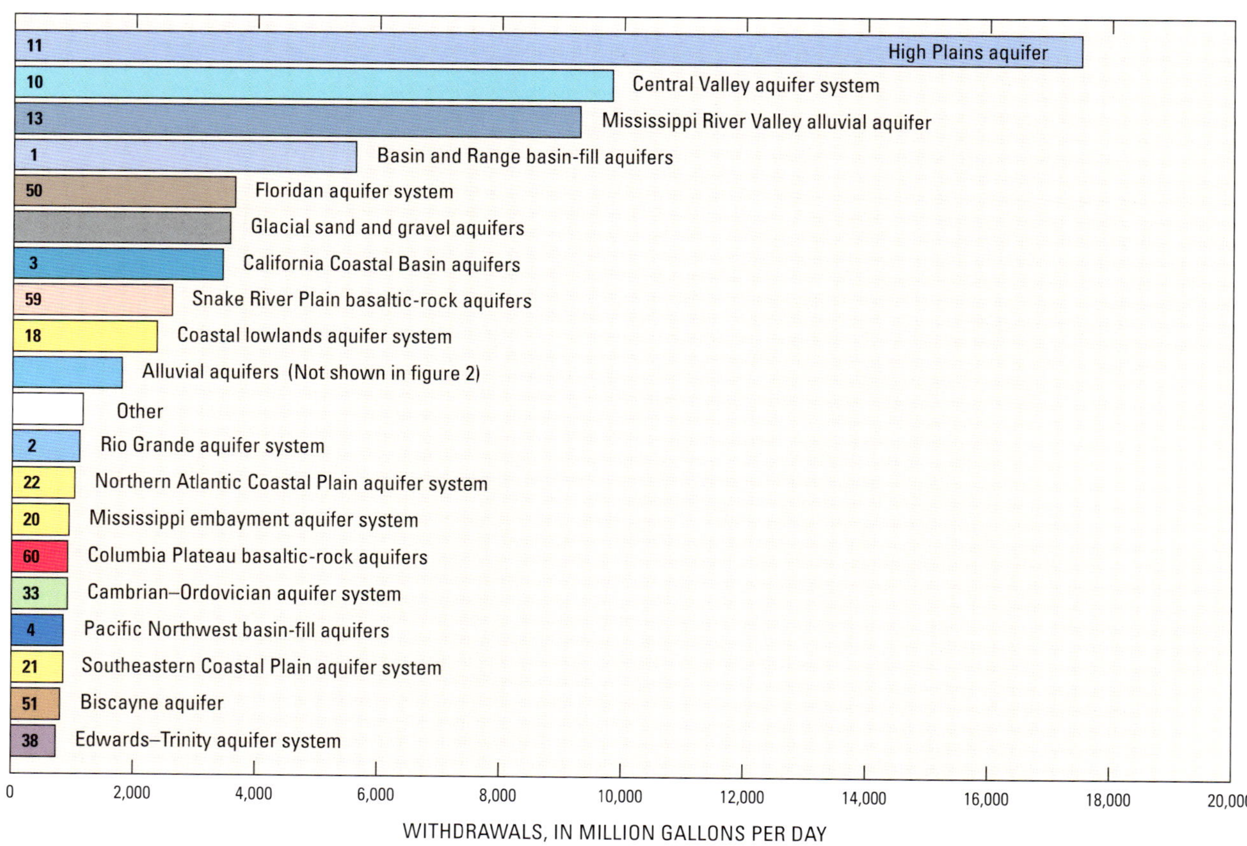

Figure 5. Principal aquifers that provided most of the total withdrawals for irrigation, public-supply, and self-supplied industrial water uses in the United States during 2000 (from Maupin and Barber, 2005). See figure 2 for aquifer locations.

Changes in Ground-Water Levels and Ground-Water Storage

The response of a ground-water system to withdrawals was explained succinctly by Theis (1940). Two important points raised by Theis (1940, p. 10) are: "All water discharged by wells is balanced by a loss of water somewhere," and "This loss is always to some extent and in many cases largely from storage in the aquifer. Some ground water is always mined." Discharge from a well always requires some drawdown (change in water level) to create hydraulic gradients to move water to the well. Thus, all discharging wells have to remove some water from storage, and all systems must conserve mass. This conservation of the mass of water is referred to as a mass balance.

The altitude of the top of the water column in a well is called the water level. Water levels in carefully constructed wells with short screens indicate the energy or head of the water and can be used to define the direction and rate of movement of the water in the ground-water system (for more information see Taylor and Alley, 2001). Water levels are important because they define the state of the saturated ground-water system. Decreases in water levels caused by pumping not only indicate changes in the amount of water stored in the aquifer but also that water probably is being "captured" from a neighboring surface source. The relative significance of water-level changes with respect to these two aspects depends on the location of the water-level changes in relation to the boundaries of the ground-water system.

In systems that have streams that are nearby and not deeply incised in the aquifer (fig. 6), a small change in water level can change the ground-water flow to the stream by a large percentage. If the water level in the ground-water system drops below the stream stage (the water level of the stream), the stream can become a major source of flow to the ground-water system. The change in the amount of water stored in the ground-water system would not change that much in this example, but the effects on the nearby stream system could be substantial. In systems that are expansive and have boundaries that are far away, such as the southern High Plains aquifer system (fig. 7), there can be large changes in both water levels and the amount of water in storage before the distant bounding streams or springs are affected.

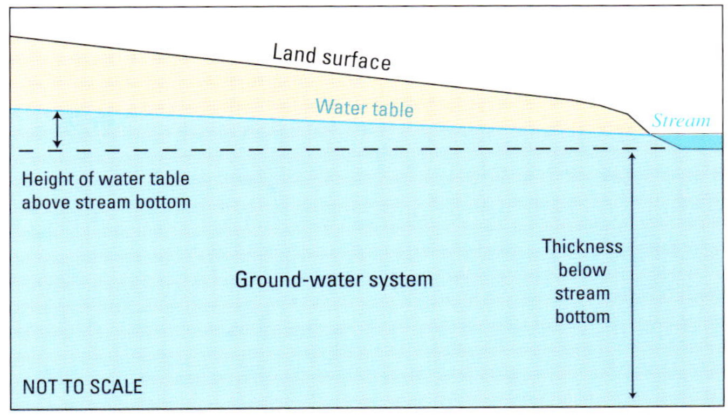

Figure 6. A cross section of a hypothetical stream-aquifer system, showing that a small change in the ground-water level can result in a large change in gradient and flow to the stream.

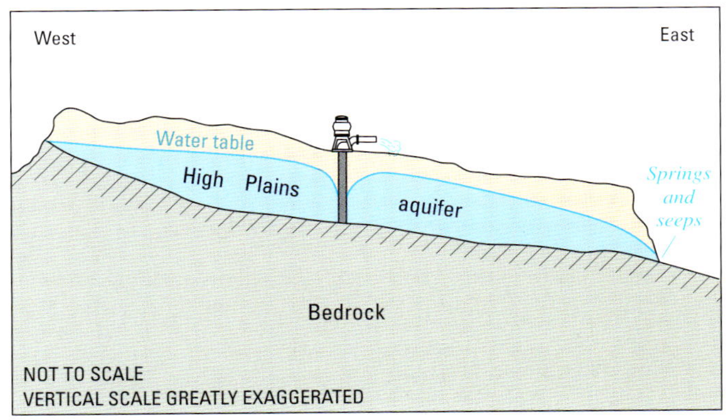

Figure 7. A diagrammatic cross section of the southern High Plains aquifer system, illustrating that in areally extensive systems, large changes in water levels and in the amount of water stored may occur before the effects of the water-level changes reach the boundaries of the system (modified from Lohman, 1972).

Ground-water systems are continuous saturated systems made up of different earth materials. As a simplified classification, these saturated earth materials can be classified as either aquifers or confining beds. As previously defined, an aquifer contains sufficient saturated permeable material to yield significant quantities of water to wells and springs. A confining bed is a rock unit of very low hydraulic conductivity that restricts the movement of ground water either into or out of adjacent aquifers (Heath, 1983). A ground-water system can be made up of many aquifers and confining beds. The top boundary of the saturated ground-water system is the water table.

The response of an aquifer through time to pumping depends in part on whether the aquifer is "confined" or "unconfined." Aquifers that contain the water table are referred to as unconfined aquifers. In unconfined aquifers, the drawdowns through time in response to pumpage usually are moderated because of the large source of water derived from storage as water is drained from the pore space of the aquifer material as the water level drops. In addition, in unconfined aquifers, surface-water sources of water can be nearby and can limit the extent of the water-level declines.

Aquifers that are fully saturated and have confining units above them are referred to as confined (or artesian) aquifers. Drawdowns through time in response to pumping confined aquifers usually are larger than those in unconfined aquifers for the same transmitting properties. The water table at the very top of the entire ground-water system is hydraulically distant from the direct influence of the pumping in a confined aquifer and hence the water-table aquifer does not drain in response to the pumping. Rather, as water levels drop in response to pumpage, the water pressure in the confined aquifer decreases, and water is released from storage as the water expands and the aquifer material compresses. In addition, in confined aquifers the boundaries of the aquifer are usually farther away physically (or hydraulically) and, thus, cannot readily serve as a source of water to be captured.

If the ground-water system eventually comes to a new equilibrium in response to the pumping, the effect of storage on the magnitude of the drawdown is no longer a factor because water levels are no longer changing, regardless if the aquifer is confined or unconfined. The magnitude of the drawdown under equilibrium conditions depends on the transmitting properties and the location and magnitude of the sources of water.

In some hydrogeologic terrains, the removal of water can cause some fine-grained confining beds to compact substantially causing subsidence of the land surface. Land subsidence can be a gradual or sudden sinking of the Earth's surface caused by subsurface movement of earth materials (Galloway and others, 1999). Some areas of the Nation, such as the San Joaquin Valley, California (fig. 8), are more prone to subsidence because of the presence of fine-grained compressible confining beds. Thus, water-level declines become very important in areas susceptible to subsidence.

Figure 8. Approximate location of maximum subsidence in United States identified by research efforts of Joseph Poland (pictured). Signs on pole show approximate altitude of land surface in 1925, 1955, and 1977. The pole is near benchmark S661 in the San Joaquin Valley southwest of Mendota, California (from Galloway and Riley, 1999).

The volume of ground water in storage is decreasing in many areas of the United States in response to withdrawals. If these water-level declines are sustained over time, the effect is often described as ground-water depletion. Among the consequences of ground-water-level declines are increased pumping costs, deterioration of water quality, reduction of water in streams and lakes, and land subsidence. Such negative effects, while variable, happen to some degree with any ground-water use. As with other natural resources, society must weigh the benefits gained by the use of the natural resource against the consequences of such use. These effects should be observed over time in order to determine their impact.

The extent of ground-water-level declines across the United States has not been monitored or computed on a regular basis. Previous compilation efforts have included national maps of "Areas where significant cones of depression have been developed by pumping from wells" and "Ground-water reservoirs with perennial overdraft" (Thomas, 1951, Plates II and III), and a national map showing the "Area in which significant ground-water overdraft is occurring" (U.S. Water Resources Council, 1978, p. 59). The most recent national summary of ground-water-level declines was compiled by the USGS in 1983. The map from this compilation is shown in figure 9, which delineates areas of water-table decline or artesian water-level decline in excess of 40 feet (ft) in at least one aquifer since predevelopment (U.S. Geological Survey, 1984).

There are inherent difficulties with a national compilation of this kind. Aquifers may be confined or unconfined, the ground-water system may be directly connected or distant from a surface-water body, and ground-water development may be uneven in space and time. Ground-water-level declines occur at scales ranging from a single well to aquifer systems underlying several States. The extent of the resulting effects depends on several parameters, including pumping and natural discharge rates, physical properties of the aquifer, and natural and human-induced recharge rates. At the local scale, monitoring of water-level declines commonly focuses on local effects (such as local well interference), and at the regional scale, monitoring commonly focuses on more widespread effects (such as streamflow depletion).

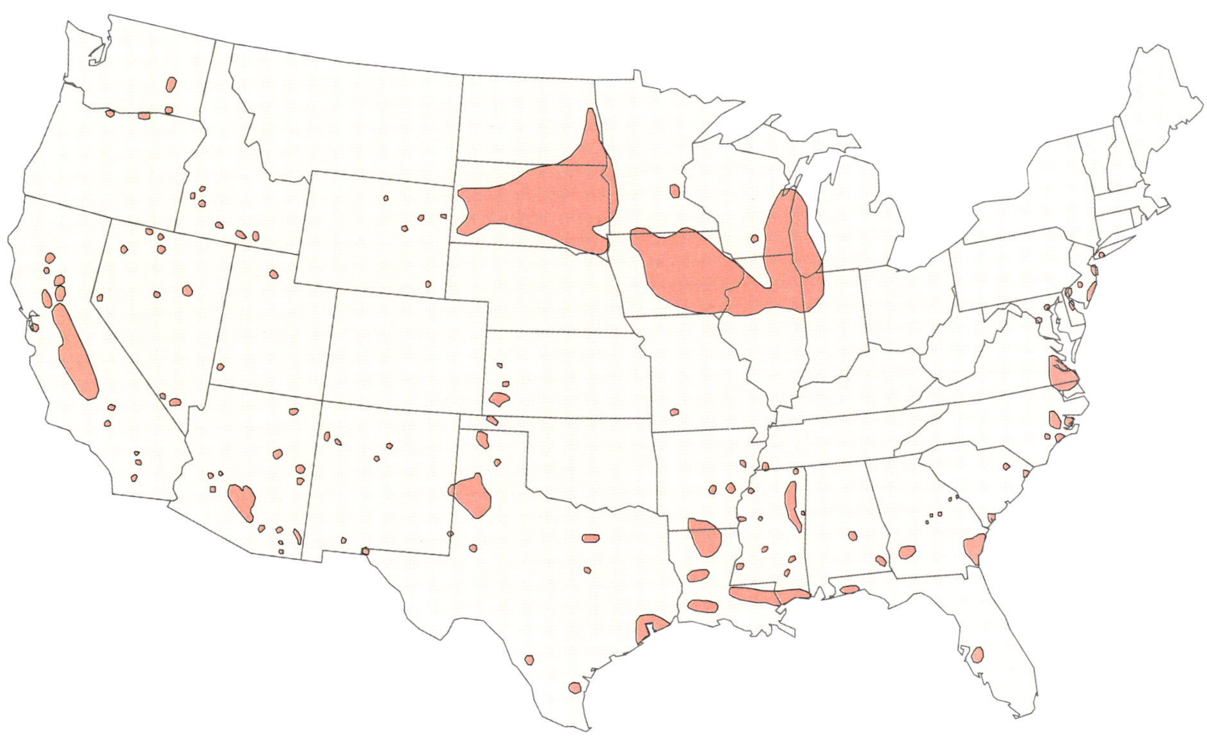

Figure 9. Areas of water-table decline or artesian water-level decline in excess of 40 feet in at least one aquifer since predevelopment (modified from U.S. Geological Survey, 1984).

To illustrate the extent of ground-water-level declines over a national and regional scale, information was gathered in 2007 from USGS State Water Science Centers throughout the Nation for areas of water-level decline in excess of 40 ft in at least one confined aquifer since predevelopment and areas of water-level decline in excess of 25 ft in an unconfined aquifer since predevelopment (fig. 10). In order to be included on the map, the areal extent of the water-level decline had to be approximately 500 square miles (mi^2) or larger. In some cases, USGS State Water Science Centers suspected additional areas of ground-water decline, but sufficient supporting data were not available to define the extent of the areas; therefore, these areas were not included on the map.

Because ground-water-level declines are necessarily the result of ground-water use, these areas of extensive ground-water-level declines correspond well to the highly used principal aquifers. A description of selected areas of ground-water-level declines follows, with reference to the principal aquifers shown in figure 2. In comparison to figure 9, some areas shown in figure 10 have expanded since the 1983 compilation. In other areas, the water-level changes took place during initial development, and water levels are relatively stable today (2007). For example, most of the area shown as declines in the upper Midwest (mainly in South Dakota and Iowa) is the result of drawdowns that occurred early in the 1900s (U.S. Geological Survey, 1984).

Although this approach to evaluate significant areas of ground-water-level decline is useful, it is not comprehensive. In some aquifers, such as the Basin and Range basin-fill aquifers, the geologic structure of the aquifer limits the areal extent of the drawdown (though it may exacerbate the rate of decline). Basin and Range basin-fill aquifers generally are composed of unconsolidated material in basins bounded by mountains. Individual basins may not be large enough to meet the criteria used in this analysis.

Because ground-water-level decline can have impacts at a variety of scales, a national map showing potential local effects also was compiled. This is not a comprehensive evaluation of water-level declines in all areas across the Nation, because a comprehensive database of all ground-water-level monitoring data in the United States does not exist. In fact, the

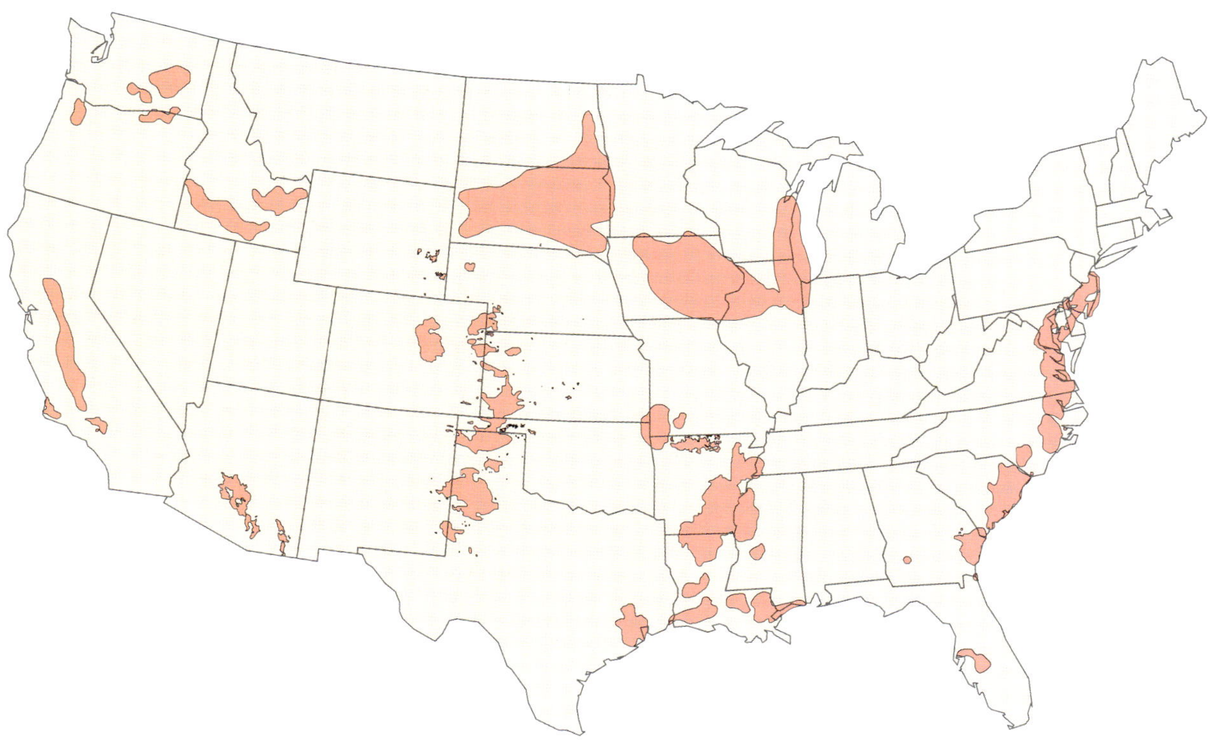

Figure 10. Areas of water-level decline in excess of 40 feet in at least one confined aquifer since predevelopment, and areas of water-table decline in excess of 25 feet in the water-table aquifer since predevelopment.

availability of ground-water levels and rates of change is "not adequate for national reporting" according to the report, "The State of the Nation's Ecosystems" (The H. John Heinz III Center for Science, Economics, and the Environment, 2002). A followup report from the Heinz Center (The H. John Heinz III Center for Science, Economics, and the Environment, 2006) identified ground-water levels as "one of the 10 highest priority data gaps that must be filled to improve the nation's ability to report on ecosystem conditions and use, and to make sound policy and operational decisions."

Even though it does not contain all the wells measured by other agencies, the USGS National Water Information System (NWIS) database contains the most complete ground-water-level dataset for the Nation. NWIS contains ground-water-level data from USGS Federal programs, programs with other Federal agencies, and cooperative programs with State and local governments. Water-level data in the NWIS database may or may not include predevelopment measurements, so the total water-level declines reported from an individual well may be underestimated. In addition, water-level changes for this analysis were determined by calculating the difference between the highest and lowest water level measured over time. This approach does not account for seasonal water-level recovery, or long-term recovery, in some cases. Figure 11 illustrates individual wells where the difference between the highest and lowest water level measured is at least 40 ft. It is clear from figure 11 that few areas of the Nation have been spared from the effects of substantial water-level declines.

Summaries of some of the areas experiencing substantial water-level declines are provided below. Other areas, including the California Central Valley aquifer system, the High Plains aquifer system, and the aquifer systems surrounding the Great Lakes, are described later in this report as case studies and are not presented here. The summaries below and in the case studies suggest that water-level declines are widespread across the Nation and can result from different circumstances.

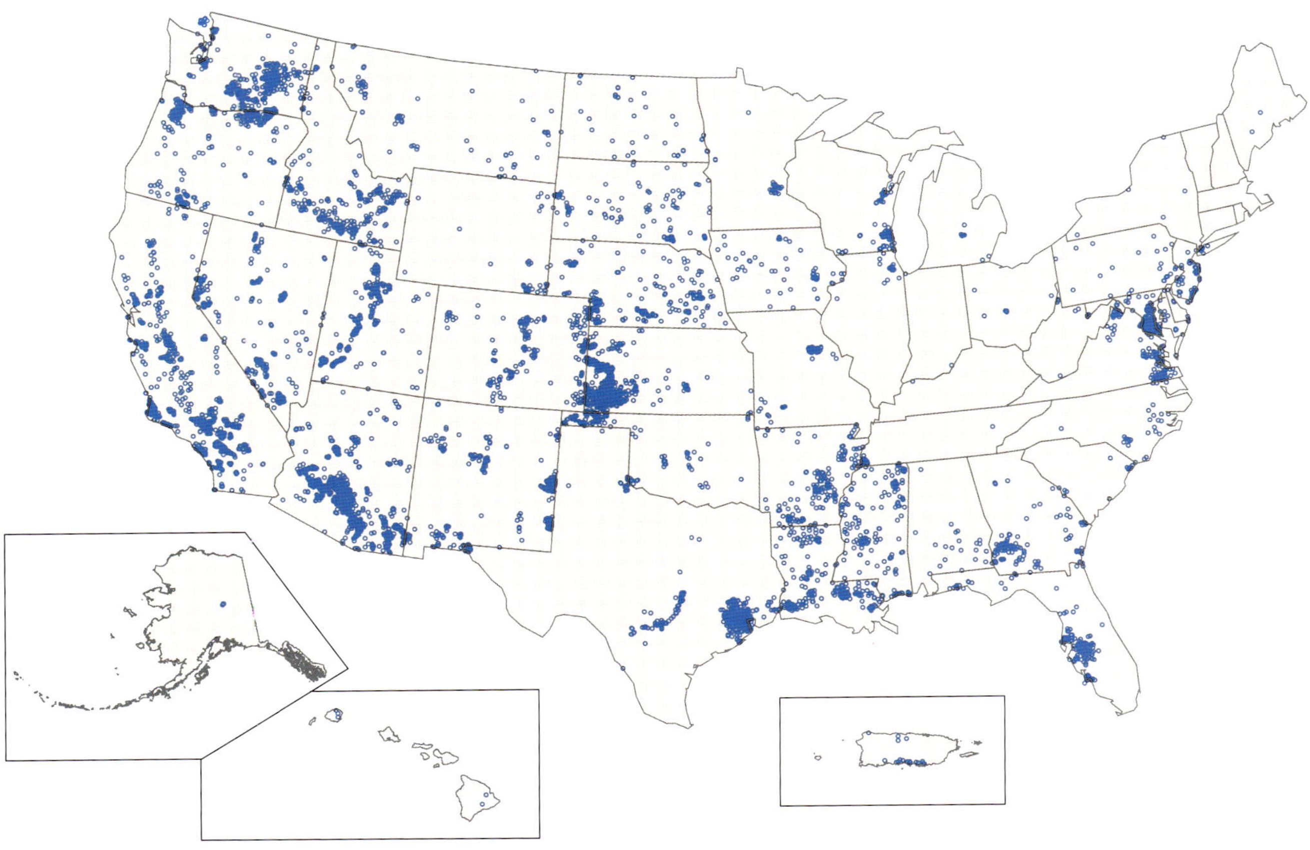

Figure 11. Individual wells in the USGS National Water Information System database where the difference between the lowest and highest water-level measurement over time is at least 40 feet. Each well is one blue circle, and apparent miscellaneous shapes are due to overlapping circles.

Pacific Northwest Region

Evaluating the effect of ground-water use on the history and extent of water-level declines in some parts of the country is complicated by the fact that excess surface-water irrigation can contribute to water-level increases or reduce declines that would have occurred otherwise. Two aquifer systems in the Pacific Northwest provide such examples.

Water use for irrigation, public supply, and industrial uses in the area of the Columbia Plateau basalt aquifer system of Washington and Oregon has resulted in significant water-level changes. Water levels have risen (more than 300 ft in some areas) to near land surface beneath about 1.2 million irrigated acres as a result of the application and infiltration of surface water. Water-level declines resulting from ground-water pumping for irrigation began in the 1960s. The declines continue today (2007) and have exceeded 200 ft in some parts of the central plateau that are not irrigated by surface water (Vaccaro, 1999).

The Snake River Plain aquifer in Idaho provides extensive water for irrigation as well as much of the flow of the Snake River through springs. Prior to about 1960, surface-water irrigation raised water levels in parts of the Snake River Plain aquifer by 60–70 ft and increased ground-water discharge to the river and springs (Idaho Department of Water Resources, 1999). Since the high water-level period of the 1950s, water levels and spring discharge have decreased because of intensive use of ground water for agriculture and more efficient irrigation practices (Lindholm, 1996). Areas of water-level decline in these two principal aquifers have increased since the 1983 compilation.

Mississippi River Valley and Gulf Coast Region

Ground-water use in the Mississippi River Valley alluvium, the Coastal lowlands, and the Mississippi embayment aquifer systems has been fundamental to the region's agriculture, industry, and some municipalities. In the Houston, Texas area, extensive pumpage of ground water to support economic and population growth has caused water-level declines of approximately 400 ft, resulting in extensive land-surface subsidence of as much as 10 ft. The City of Houston is addressing this problem by shifting to surface-water sources as their primary supply. Ground-water pumpage for Baton Rouge, Louisiana, increased more than tenfold between the 1930s and 1970, resulting in ground-water-level declines of approximately 200 ft. Baton Rouge is underlain by a series of aquifers, and as one aquifer experiences large water-level declines, pumping has shifted to others over time. Ground-water declines in the Sparta aquifer in Arkansas, Louisiana, Mississippi, and Tennessee have raised concerns about the sustainability of the resource. Some areas have shifted their source of supply to surface water. Regional water-level declines of as much as 70 ft have resulted in interstate concerns over continued and increased pumpage in the Memphis, Tennessee area.

Atlantic Coastal Region

Although the humid eastern seaboard has abundant renewable freshwater resources, ground-water resources in the highly productive principal aquifers are under increasing stress in some areas as coastal populations continue to increase. This increase in ground-water use from the Floridan, Northern Atlantic Coastal Plain, and Southeastern Coastal Plain aquifer systems has resulted in a significant expansion of the areas affected by ground-water declines since the 1983 compilation. The aquifers of the Atlantic Coastal Region generally are confined aquifers, with productive units commonly overlying one another. The area depicted for the Atlantic Coastal Region in figure 10 consists of 45 individual areas of drawdown calculated using water levels from predevelopment and the year 2000 that have been merged together for this map (dePaul and others, 2008).

Basin and Range Aquifers

Ground-water-level declines in Basin and Range aquifers in California, Arizona, Nevada, and Utah become observable when examining the declines from individual wells. Ground-water use from Basin and Range aquifers is the fourth highest among the Nation's principal aquifers. Water is needed in the area to support a fast-growing population as well as agricultural production in the desert southwest. Drawdown patterns from ground-water use can be seen in southern Arizona, southern California, the Las Vegas, Nevada area, and central Utah.

The information presented in figure 12 combines the regional water-level declines presented in figure 10 with the more local water-level declines presented in figure 11. Figure 12 represents the current state of knowledge about ground-water declines on a national scale. Our knowledge is incomplete in some areas because there are not enough water-level data and perhaps in other areas because data have not been compiled. A national effort is needed to organize available information on changes in ground-water storage, similar to what was done for the High Plains aquifer (see High Plains aquifer case study). To address this need, the Advisory Committee on Water Information in 2007 created a Subcommittee on Ground Water (*http://acwi.gov/sogw/index.html*) to develop and encourage implementation of a nationwide, long-term ground-water quantity and quality monitoring framework.

When the necessary data do not exist, additional monitoring is needed to refine the knowledge of water-level declines on a national scale. Moreover, observation of ground-water-level changes is only one piece of the puzzle concerning ground-water availability. Measurements that indicate changes in base flow of streams are also good indicators of changes in ground-water levels. In addition, water-level declines occur in three dimensions, and differences in changes with depth are important although they cannot be shown on the two-dimensional national maps.

New methods and scientific advances can continue to improve our understanding and ability to document water-level and storage changes. For example, the use of microgravity measurements on the ground (Pool and Eychaner, 1995) and from space (Swenson and others, 2003) is being developed and refined to estimate storage changes. These methods for estimating changes in ground-water storage are promising at the local and regional scales.

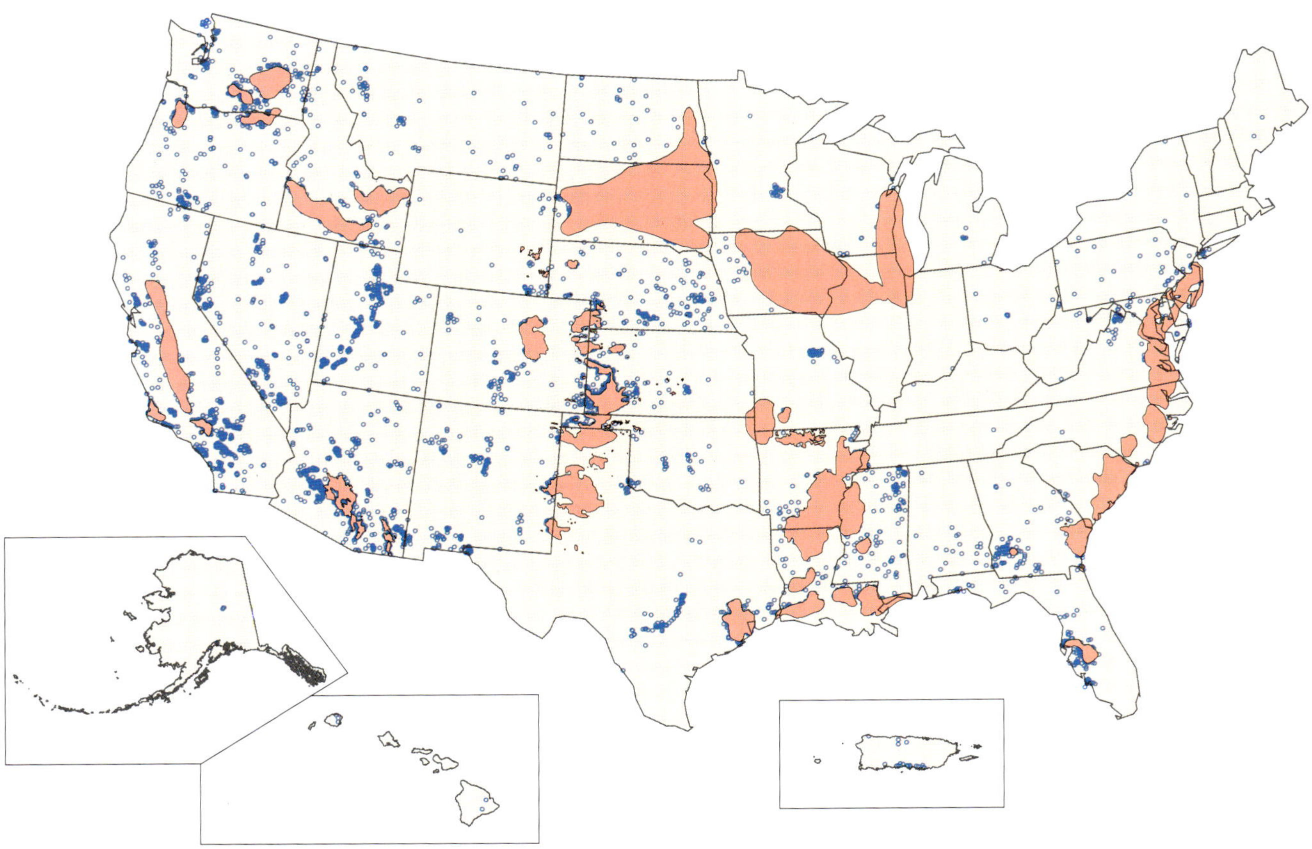

Figure 12. Water-level declines. Red regions indicate areas in excess of 500 square miles that have water-level decline in excess of 40 feet in at least one confined aquifer since predevelopment, or in excess of 25 feet of decline in unconfined aquifers since predevelopment. Blue dots are wells in the USGS National Water Information System database where the measured water-level difference over time is equal to or greater than 40 feet.

Recharge

Estimates of recharge quantify the amount of water entering the saturated ground-water system. These estimates, however, are dependent on the scale of the estimate, the method used to make the estimate, and the time over which the estimate is made. The intended use of the estimated recharge value is important in determining the scale, time period, and accuracy needed. The following discussion is limited to recharge to the water table (as opposed to interaquifer recharge).

Recharge can be diffuse or localized. Diffuse recharge refers to the widespread movement of water from land surface to the water table as a result of infiltration of precipitation over large areas and percolation through the unsaturated zone. Localized recharge, on the other hand, refers to the movement of water from surface-water bodies to the ground-water system and is less uniform in space than diffuse recharge. Most ground-water systems receive both diffuse and localized recharge. In general, the importance of diffuse recharge decreases as the aridity of a region increases. Information available on recharge at the national scale pertains to diffuse recharge. The information available does not account for the effects of irrigation and other human activities, which can have major effects on recharge in many areas.

The amount of precipitation represents the maximum amount available (an upper limit) of possible natural diffuse recharge to the ground-water system. Precipitation frequently is the principal source of water to the ground-water system in the absence of irrigation. Examination of an average precipitation map for the United States (fig. 13) is useful in determining broad areas of potential high and low recharge.

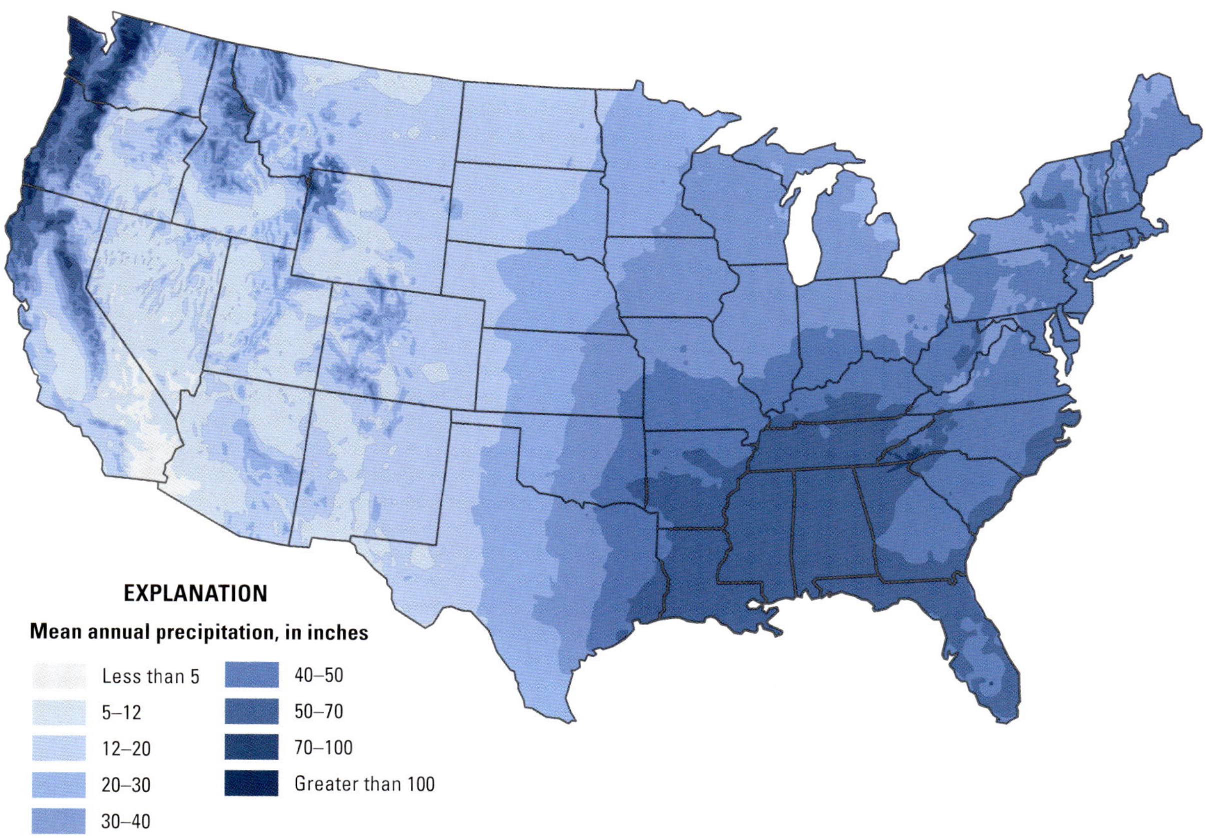

Figure 13. Mean annual precipitation for the conterminous United States, 1890 to 2002 (from Anderson and Woosley, 2005).

The amount of water available for natural recharge to the ground-water system and as surface runoff to streams is represented by the amount of precipitation minus the amount of evapotranspiration. Evapotranspiration is the water lost to the atmosphere by two processes: evaporation and transpiration by plants. In the total water budget for an area, evapotranspiration is a significant budget component. It is estimated that evapotranspiration for the conterminous United States accounts for 67 percent of the outflow of water from precipitation (Hanson, 1991). In arid areas, evapotranspiration often exceeds precipitation in the long term, and areal recharge occurs only sporadically following extreme rainfall events. Evapotranspiration amounts are difficult to measure and are usually estimated.

Figure 14 is an indicator of potential natural recharge on a national scale. The map of the conterminous United States in figure 14 shows the distribution of the difference between precipitation and potential evapotranspiration (Healy and others, 2007). Potential evapotranspiration is "the water loss that would occur by evapotranspiration if there was never a deficiency of water in the soil for use by vegetation" (Wilson and Moore, 1998). Where precipitation substantially exceeds potential evapotranspiration, one can expect a water surplus, and where potential evapotranspiration substantially exceeds precipitation (the negative numbers), one can expect a water deficit. The map indicates that the eastern United States has more precipitation than potential evapotranspiration and the Great Plains and southwestern United States has more potential evapotranspiration than precipitation.

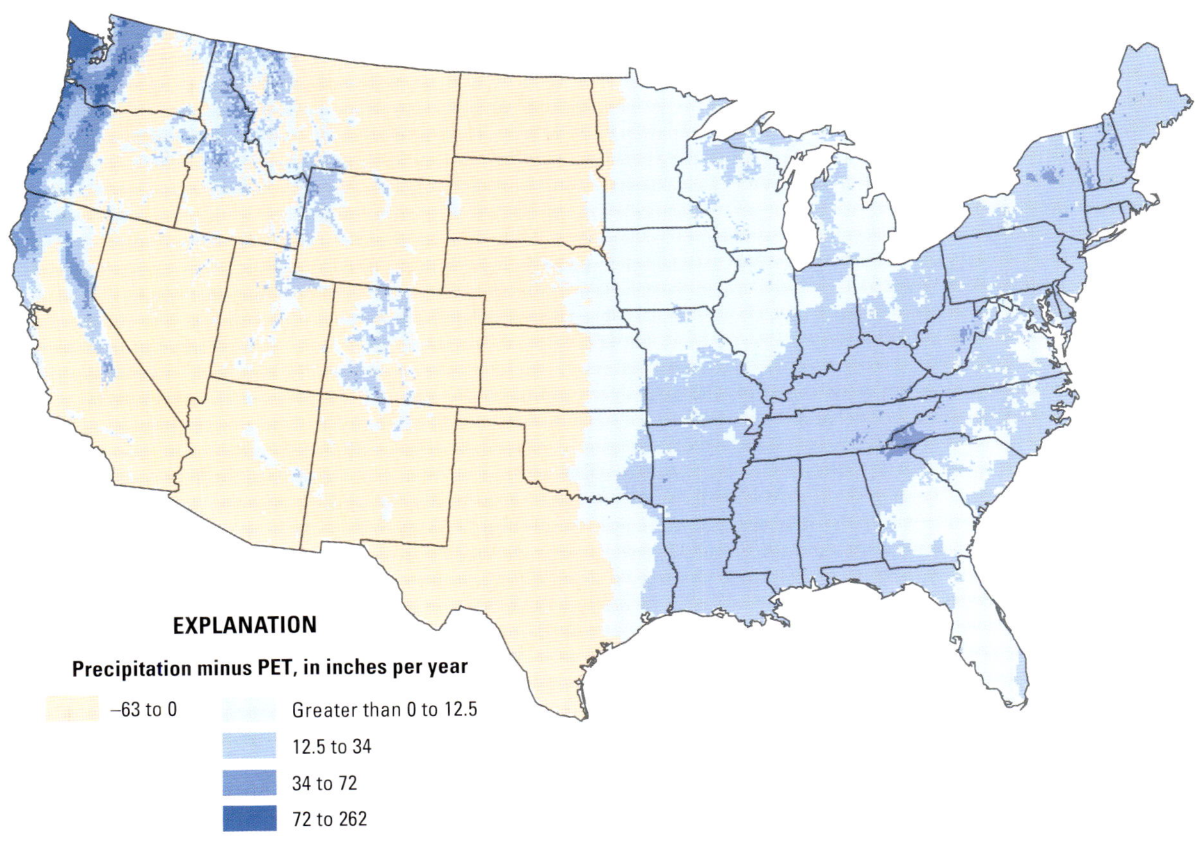

Figure 14. Difference between annual precipitation and potential evapotranspiration (PET) rates across the conterminous United States (modified from Healy and others, 2007).

Figure 15 (Roy and others, 2005) takes the analysis shown in figure 14 one step further and estimates the precipitation available for use in the United States; that is, how much water is available for recharge to the ground-water system or as runoff to streams. This estimate was calculated as the difference between precipitation and potential evapotranspiration summed for all months in the year in which precipitation exceeded potential evapotranspiration. The map in figure 15 shows values of precipitation available for use at the county level estimated from data from 1934 to 2002 at the 344 National Oceanic and Atmospheric Administration climate divisions covering the conterminous United States. This map

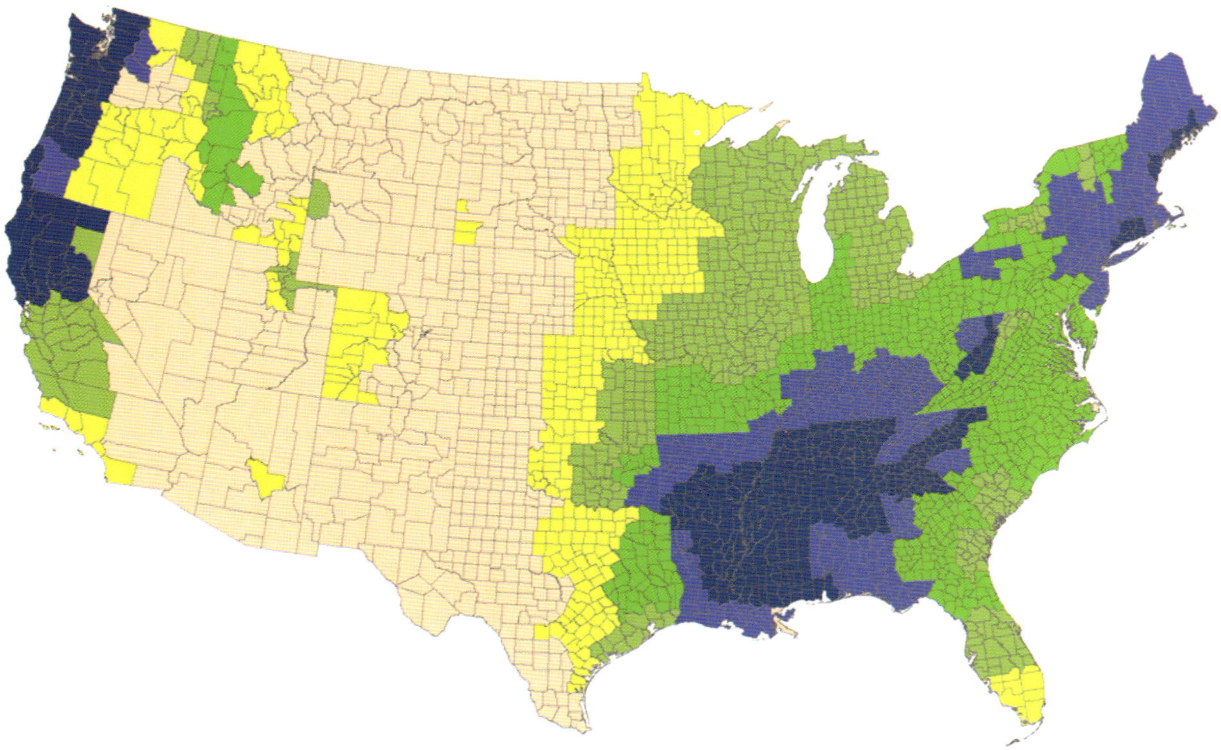

EXPLANATION

Available precipitation—Average from 1934 to 2002, in inches per year. Number of counties in parentheses

- 0 to 5 (603)
- 5 to 10 (414)
- 10 to 15 (525)
- 15 to 20 (698)
- 20 to 25 (464)
- Greater than or equal to 25 (407)

Figure 15. Available precipitation (difference between monthly precipitation and potential evapotranspiration [PET], sum of months with nonzero values) for the conterminous United States (modified from Roy and others, 2005; reprinted with permission, American Water Resources Association).

clearly shows that most of the western United States, except for some coastal areas, has far less water available for ground-water recharge and use than the rest of the country. Thornthwaite (1948) presented a similar map (fig. 16) that showed qualitative areas of water deficiency across the conterminous United States. Figure 16 shows areas of little or no water deficiency, areas of seasonal moisture variation (summer deficiency and winter surplus), and areas with little or no water surplus. The maps by Healy and others (2007), Roy and others (2005), and Thornthwaite (1948) describe a similar assessment of the distribution of water available for recharge and runoff across the United States.

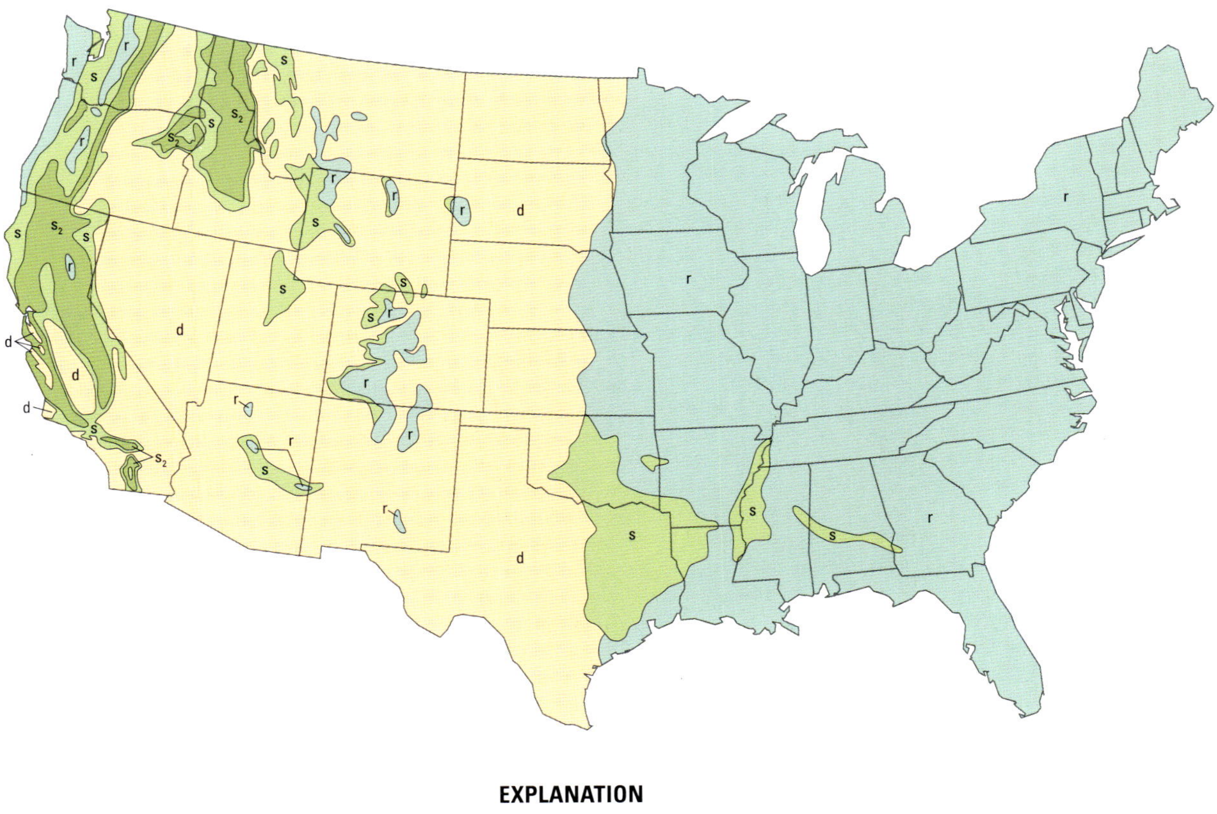

EXPLANATION

- r — **Little or no water deficiency in any season**—Occurs only in moist climates
- s — **Moderate seasonal moisture variation with summer the driest season**—Summer water deficiency in moist climates; winter water surplus in dry climates
- s_2 — **Large seasonal moisture variation with summer the driest season**—Summer water deficiency in moist climates; winter water surplus in dry climates
- d — **Little or no water surplus in any season**—Occurs only in dry climates

Figure 16. Seasonal variation of effective moisture for the conterminous United States (modified from Thornthwaite, 1948; reprinted with permission, The American Geographic Society).

Estimated values of natural ground-water recharge are available for the entire United States (Wolock, 2003a) based on base-flow separation techniques. The component of streamflow that is contributed by ground-water discharge as opposed to surface runoff is called base flow. Wolock (2003a) produced a 1-kilometer resolution raster dataset (fig. 17) as an estimate of mean annual natural ground-water recharge. The dataset was created by multiplying a grid of base-flow index values (Wolock, 2003b) by a grid of mean annual runoff values derived from a 1951–80 mean annual runoff contour map (Gebert and others, 1987). The concept used to construct the estimate is based on two assumptions: (1) long-term average natural ground-water recharge is equal to long-term average natural ground-water discharge to streams, and (2) the base-flow index reasonably represents, over the long term, the percentage of natural ground-water discharge in streamflow. These estimates (fig. 17), based on very broad assumptions, give a "big picture" of natural ground-water recharge but cannot characterize the variability over space and time at more local scales. In addition, in areas where streams are regulated, irrigation return flows are large, or urbanization has occurred, the basic assumptions may not hold. For example, three of the case studies presented later in this report (Middle Rio Grande Basin, California Central Valley aquifer system, and the High Plains aquifer) illustrate several-fold increases in recharge from natural to developed conditions.

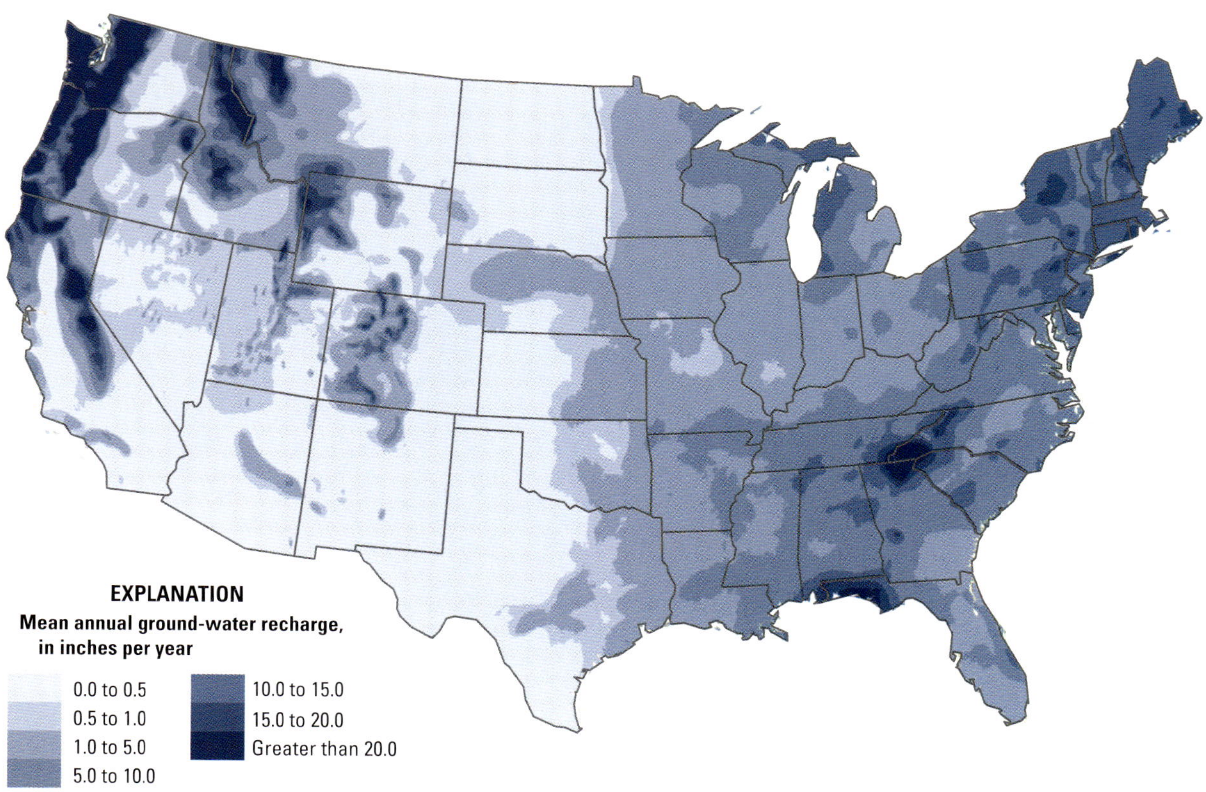

Figure 17. Estimated mean annual natural ground-water recharge in the conterminous United States (data from Wolock, 2003a).

There are estimates of recharge for large parts of the United States, such as the Great Lakes Basin (Neff and others, 2006), and there are estimates for smaller areas (Gebert and others, 2006). Gebert and others (2006) illustrate (fig. 18) that estimates of recharge at different scales and times can show marked differences in variability even though similar base-flow separation techniques were used in all cases. In figure 18, the regional estimate produces only one value for the entire Door–Kewaunee watershed in Wisconsin, whereas the more detailed subregional examination shows a wide range of values. Gebert and others (2006) also note that the range in recharge can be expected to increase as the spatial scale becomes smaller, or different techniques are used. The objectives of a study will determine which estimates are sufficient.

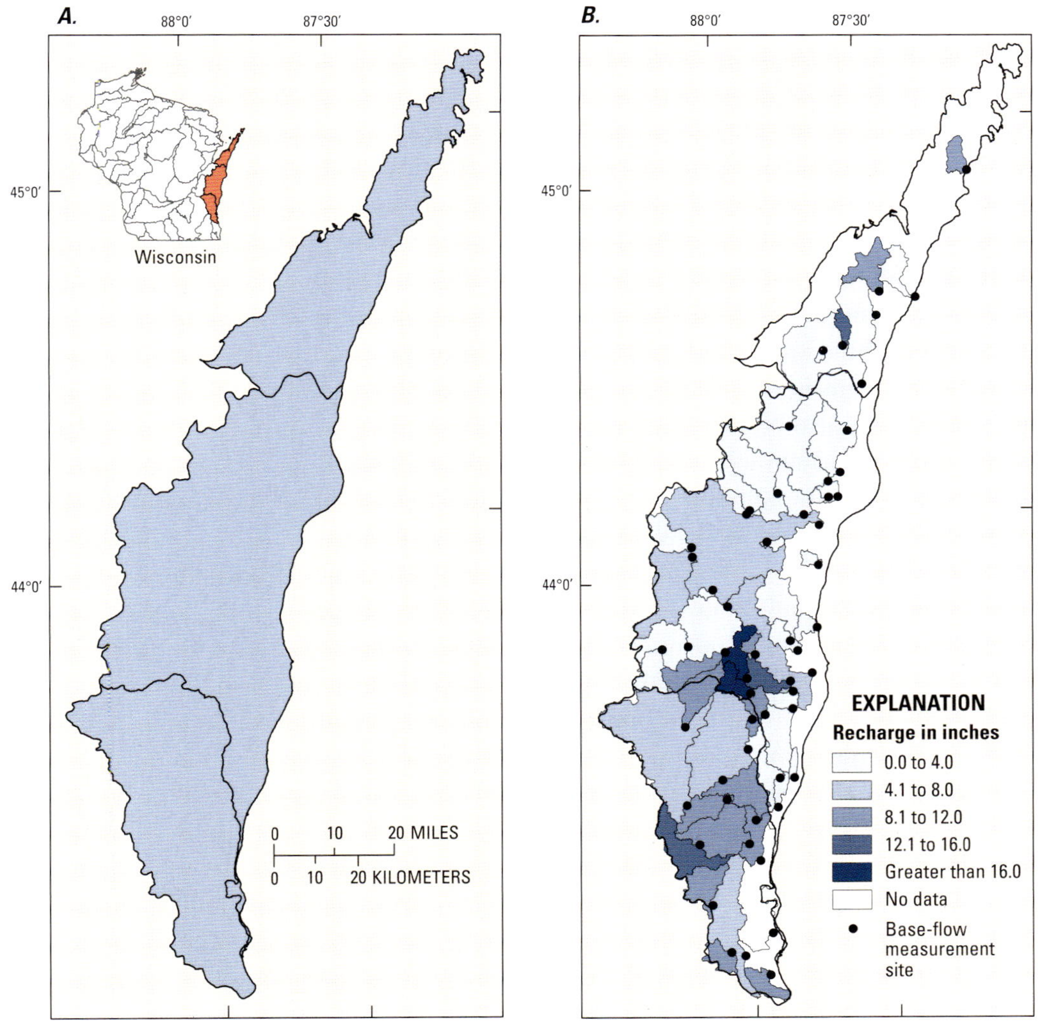

Figure 18. Recharge rates in Hydrologic Unit Code (HUC) watershed 04030102, Door–Kewaunee, estimated for two different scales *(A)* regionally and *(B)* subregionally (modified from Gebert and others, 2006).

Ground-Water Discharge

Understanding the amounts of water and where the water enters and leaves the saturated ground-water system is a prerequisite to evaluating ground-water availability. Water can naturally discharge from ground-water systems to surface-water bodies or to plants and the atmosphere. In addition, withdrawals by people also account for discharge leaving the system, as discussed previously.

Evapotranspiration directly from the ground-water system (as opposed to evapotranspiration at land surface) is mostly due to transpiration because direct evaporation from the water table through the unsaturated zone is limited. Most of the ground-water evapotranspiration takes place where the water table is close to the land surface and plant roots have easy access to the ground water. This condition usually is found in riparian zones near streams. Ground-water evapotranspiration can be a major discharge in ground-water systems in the southwestern United States but usually is not the major component of discharge in the humid eastern United States (Healy and others, 2007, p. 28). There are no national compilations of ground-water evapotranspiration, because it is highly variable and locally dependent on the depth to water and the vegetation present.

Ground-water discharge to surface-water bodies is a key component of most ground-water systems. Base flow is less variable over time than surface runoff and is responsible for keeping many streams flowing when there is no precipitation, even in times of drought. The percentage of streamflow that is accounted for by base flow for any particular stream is variable across the Nation. Wolock (2003b) estimated the ratio of base flow to total flow, expressed as a percentage (which is called the base-flow index) for the conterminous United States.

The ground-water contribution to streamflow is an important aspect of the water budget of a ground-water system because it frequently is the primary source of water that can be captured for use. Understanding the changes in the discharge of a ground-water system to development through time is fundamental to understanding ground-water availability (Bredehoeft, 2007).

Ground-Water Quality

The chemical quality of ground water affects its suitability for different uses. Natural chemical reactions between the water and the rock it flows through add dissolved substances and compounds to the water. Human-made contaminants at the land surface also can percolate into ground water and affect its suitability. The sources of ground-water contamination are numerous and diverse and include point and nonpoint (dispersed) sources. Contamination from point sources is particularly difficult to characterize from a national or regional perspective.

A basic suitability criterion is the amount of total dissolved solids or salinity of the water. Feth and others (1965) identified the depth to saline water for much of the United States (fig. 19). This national information, however, provides only a rough idea as to where salinity is an issue; the depth to saline water at any particular well can be substantially different from what is shown on the national map.

Saline water in coastal as well as inland areas has the potential to move into a fresh ground-water system and affect its suitability. Freshwater is less dense than saline water and tends to flow on top of the surrounding or underlying saline ground water. Under natural conditions, the boundary between freshwater and saltwater maintains a stable equilibrium (fig. 20). The boundary typically is not sharp and distinct but is a gradation from fresh to saline water shown by the transition zone in figure 20. When water is pumped from an aquifer that contains or is near saline ground water, the saltwater/freshwater boundary will move in response to this pumping. If the boundary moves far enough, some wells become saline, thus contaminating the water supply. The location and magnitude of the ground-water withdrawals with respect to the location of the saltwater determine how quickly and how far the saltwater moves. Even if the lateral regional movement of saltwater is negligible, individual wells located near the saltwater/freshwater boundary can become saline as a result of significant local drawdowns that cause underlying saltwater to "upcone" into the well.

The availability of ground water and the suitability of its quality for different uses are inextricably intertwined. As just described, saltwater occurs

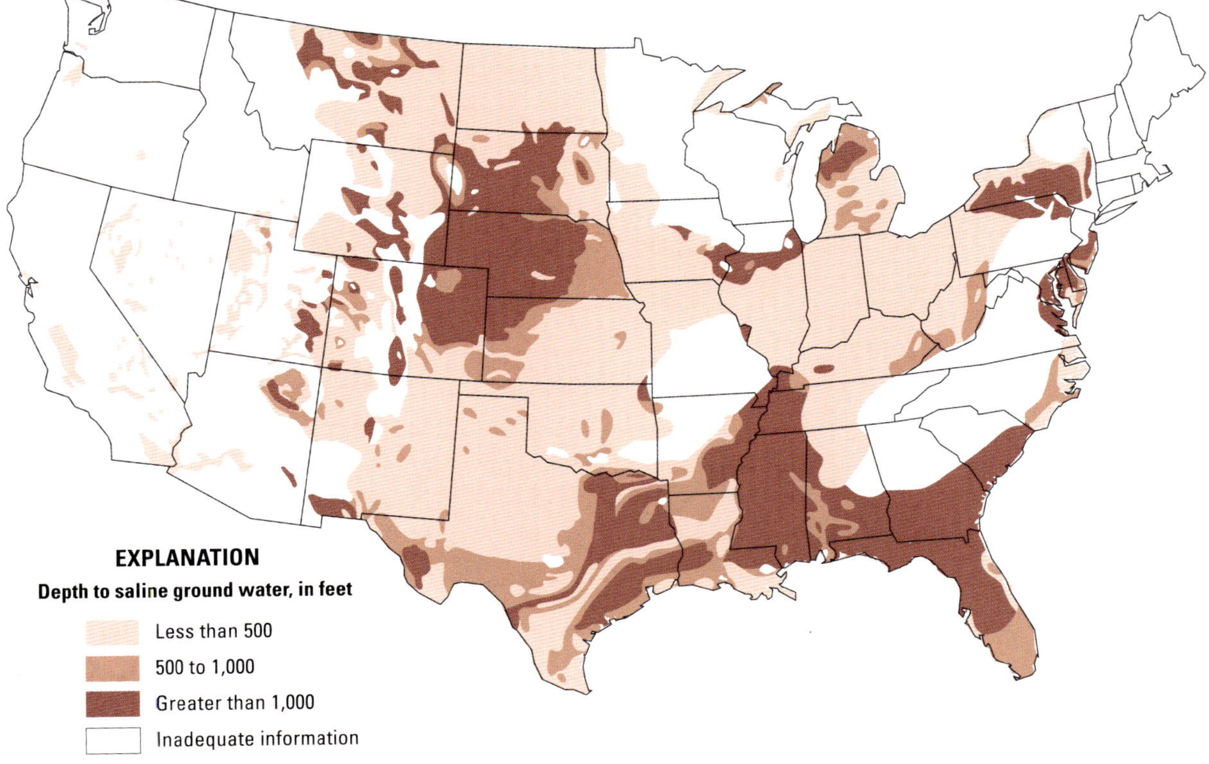

Figure 19. Depth to saline ground water in the conterminous United States (generalized from Feth and others, 1965).

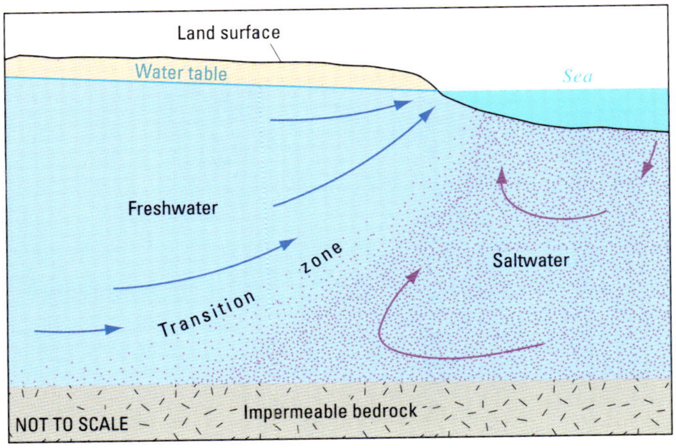

Figure 20. Ground-water flow patterns and the freshwater-saltwater transition zone in an idealized coastal aquifer. Circulation of saltwater from the sea to the transition zone and then back to the sea is induced by freshwater and saltwater mixing in the transition zone (from Barlow, 2003; modified from Cooper, 1964).

in the subsurface nationwide. Although saltwater represents huge volumes of ground water in storage, these volumes are not included in most inventories of available ground water because of their inherent unsuitability for almost all uses without treatment. Ground water of lower dissolved-solids concentrations, however, may still be suitable for some uses. For example, some cattle can tolerate a higher dissolved-solids concentration in their drinking water than can humans.

Economics is an additional factor to consider in determining the chemical suitability of a water supply. Water can be treated to remove most chemicals at a cost. For example, saltwater can be desalinated and used for drinking. Thus, when considering the chemical suitability of a source of water, one must determine the sources of all water types and the amount of money that people are willing to spend to treat it for various requirements. Likewise, there may be other issues, such as the cost of energy and environmental effects of waste disposal, when considering desalination.

The quality of ground water varies in three dimensions (areally and with depth). Some natural constituents, such as dissolved-solids concentrations, can be mapped broadly on a national scale, for example the depth to saline water as shown in figure 19. Other natural constituents, such as arsenic (fig. 21), are associated with geologic formations. Although difficult to map on a national scale because of local chemical variability, the distribution of a concentration of a particular constituent in water from wells does provide some ability to highlight areas of the Nation that may have potential problems.

Many of the chemical constituents that affect the suitability of water for use, however, come from point and nonpoint sources of anthropogenic contamination near or at the land surface. Contamination from point sources is extremely variable over a range of distances and depths and is difficult to portray at a national scale. Likewise, contamination from nonpoint sources at the land surface, such as pesticide use for crops or onsite wastewater-disposal systems for residential areas, usually is relatively local when considering the Nation and varies considerably with depth even at the local scale.

National- and watershed-scale information is available on some constituents through State and Federal programs such as the USGS National Water-Quality Assessment (NAWQA) Program (see Box B). Because of their variable spatial distribution, the information available at a national and watershed scale for many constituents consists largely of statistical information about their concentrations, frequency of occurrence, and associations with variables such as land use, chemical application rates, and other causative factors. Information from the NAWQA Program is available on a study-unit basis and as national overviews. National overviews include selected findings from 1991 to 2001 (Hamilton and others, 2004), pesticides (Gilliom and others, 2006), and volatile organic compounds (Zogorski and others, 2006).

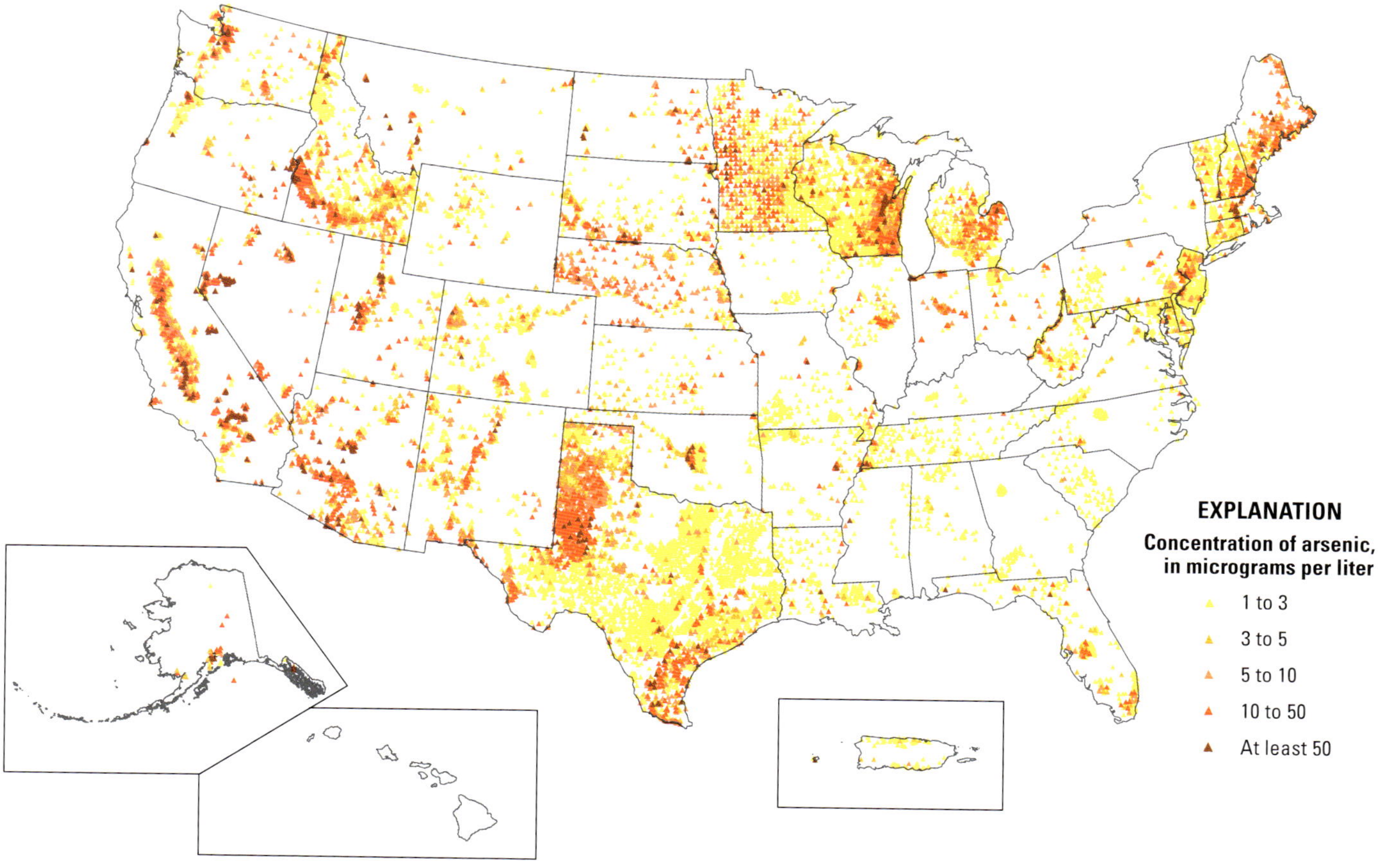

Figure 21. Ground-water arsenic samples collected in 1973–2001 (from Ryker, 2001, accessed on July 17, 2006, at *http://water.usgs.gov/nawqa/trace/pubs/geo_v46n11/fig1.html*).

B

The U.S. Geological Survey National Water-Quality Assessment (NAWQA) Program

The USGS implemented the National Water-Quality Assessment (NAWQA) Program in 1991 to develop long-term consistent and comparable information on streams, rivers, ground water, and aquatic systems in support of national, regional, State, and local information needs and decisions related to water-quality management and policy. The NAWQA Program is designed to determine the condition of our Nation's streams, rivers, and ground water, and the changes in these conditions over time.

The program collects and interprets data about surface- and ground-water chemistry, hydrology, land use, stream habitat, and aquatic life in the United States by using a nationally consistent study design and uniform methods of sampling analysis. From 1991 to 2001, the NAWQA Program conducted interdisciplinary assessments and established a baseline understanding of water-quality conditions in 51 of the Nation's river basins and aquifers (Hamilton and others, 2004).

NAWQA activities during the second decade (2001–2012) of the program are focusing in large part on national and regional assessments, all of which build on continued monitoring and assessments of the study areas. Selected major activities during the second decade include continuing national syntheses of information on pesticides, volatile organic compounds, nutrients, selected trace elements, and aquatic ecology; studies on five national priority topics, including the fate of agricultural chemicals, effects of urbanization on stream ecosystems, bioaccumulation of mercury in stream ecosystems, effects of nutrient enrichment on aquatic ecosystems, and transport of contaminants to public-supply wells; and regional assessments of water quality and trends in major river basins and principal aquifers. Additional information on the NAWQA Program is available online at *http://water.usgs.gov/nawqa/* and in Gilliom and others (2001).

Regional-Scale Approach to National Assessment

Thus far, the discussion has centered on the availability and limitations of national-scale information about the ground-water resources of the United States. As noted, the primary issues affecting ground-water availability vary from location to location and commonly require analysis in the context of ground-water flow systems to achieve a meaningful perspective. With this principle in mind, the remainder of this report focuses on how regional evaluations of the Nation's principal aquifers could form the foundation for a national assessment of ground-water availability. The information obtained on ground-water systems achieved through these regional studies, complemented by additional ground-water information available from other USGS programs, other Federal agencies, States, Tribes, and local governments could form the basis for developing a more complete picture of the Nation's ground-water availability. Regional aquifer studies currently underway in the USGS Ground-Water Resources Program (Dennehy, 2005) form the initial stages of this proposed effort.

> *"We do not have an adequate picture of water availability at national, regional, and local levels."*
>
> From "Science and Technology to Support Fresh Water Availability in the United States," National Science and Technology Council (2004)

Regional Ground-Water Budgets

Regional ground-water budgets are key to understanding the sources of water to a ground-water system and how water withdrawals change the components of flows in the hydrologic cycle. A water budget quantitatively accounts for inflows, outflows, and changes in storage of a hydrologic system. A regional ground-water budget is based on the conservation of mass within a defined regional ground-water flow system. That is, the amount of water entering the ground-water system plus the amount being removed from storage must balance the amount of water leaving the system over the time scale of interest. For further discussion of water budgets, see Healy and others (2007). In the case studies described in this report, volumetric fluxes, estimated as the average rates of change in cubic feet per second, are used to quantify the flow of water for the specified systems.

Each ground-water system in the United States is unique in terms of climate, hydrogeologic framework, and boundary conditions (both type and location), and each system responds differently to stress. Thus, the sources of water (that is, the location and magnitudes of changes in inflows, outflows, and storage) that supply withdrawals from major aquifer systems in the United States are highly variable. This variability is shown in figure 22 by the results from model simulations of the USGS Regional Aquifer-System Analysis (RASA) Program. The Floridan and Edwards–Trinity aquifer systems, which equilibrate rapidly after pumping, were simulated as steady state with no long-term change in storage. In contrast, the Southern High Plains, with most natural discharge occurring far from pumping wells, and the deeply buried Great Plains aquifer system have had significant changes in ground-water storage. The distinction between changes in recharge and changes in discharge shown in figure 22 is, in part, a function of how the system is defined (that is, a gain to one system may result in a loss from an adjoining system). For example, ground-water withdrawals from confined aquifers (such as the Northern Atlantic and Gulf Coastal Plains) can cause flow to be diverted from shallow aquifers into the deeper regional flow regime (more recharge from the perspective of the confined aquifer system) that otherwise would discharge to streams in the outcrop areas (less discharge from the perspective of the entire ground-water system). Ground-water recharge in a region also can increase as a result of human modifications, such as return flow of excess irrigation water, which has occurred in the California Central Valley. Further examples of the diversity of responses to ground-water development are given in the case studies that follow.

Selection of Regional Ground-Water Flow Systems

Many different regional ground-water flow systems can be defined for the United States. One approach for selection of regional ground-water flow systems to be studied focuses attention on the primary areas of ground-water use nationwide. This approach identifies a set of principal aquifers primarily on the basis of water use. This set of regional ground-water flow systems could be supplemented by additional aquifers in hydrogeologic terrains that may be underrepresented by the principal aquifers described and that are important current or future sources of ground water.

Figure 23 shows the set of 30 principal aquifers that collectively account for about 94 percent of the Nation's total ground-water withdrawals for public supply, irrigation, and self-supplied industrial uses combined. The principal aquifers shown in figure 23 include both individual principal aquifers and combinations of principal aquifers (where these form connected hydrologic systems) identified in the Ground Water Atlas of the United States (Miller, 2000). Figure 23 was derived by ranking the principal aquifers using the water-use data from Maupin and Barber (2005). Table 1 lists the estimated water withdrawals for 2000 (the last year these estimates were available) for the 30 principal aquifers shown in figure 23. The top ranked aquifers by water use are dominated by irrigation withdrawals. In addition to covering a substantial part of the Nation's ground-water use, the set of 30 principal aquifers covers a large spatial area of the Nation, includes a variety of hydrogeologic terrains, and includes aquifers being used to provide substantial amounts of public water supply.

Each of these 30 principal aquifers has been studied previously and to varying degrees of comprehensiveness. The Ground Water Atlas of the United States (Miller, 2000) describes all of the systems. Many were studied as part of the RASA Program from 1978 to 1994 (see Box C). Table 2 lists the 30 principal aquifers, the years they were studied as part of the RASA Program, if applicable, and more recent studies.

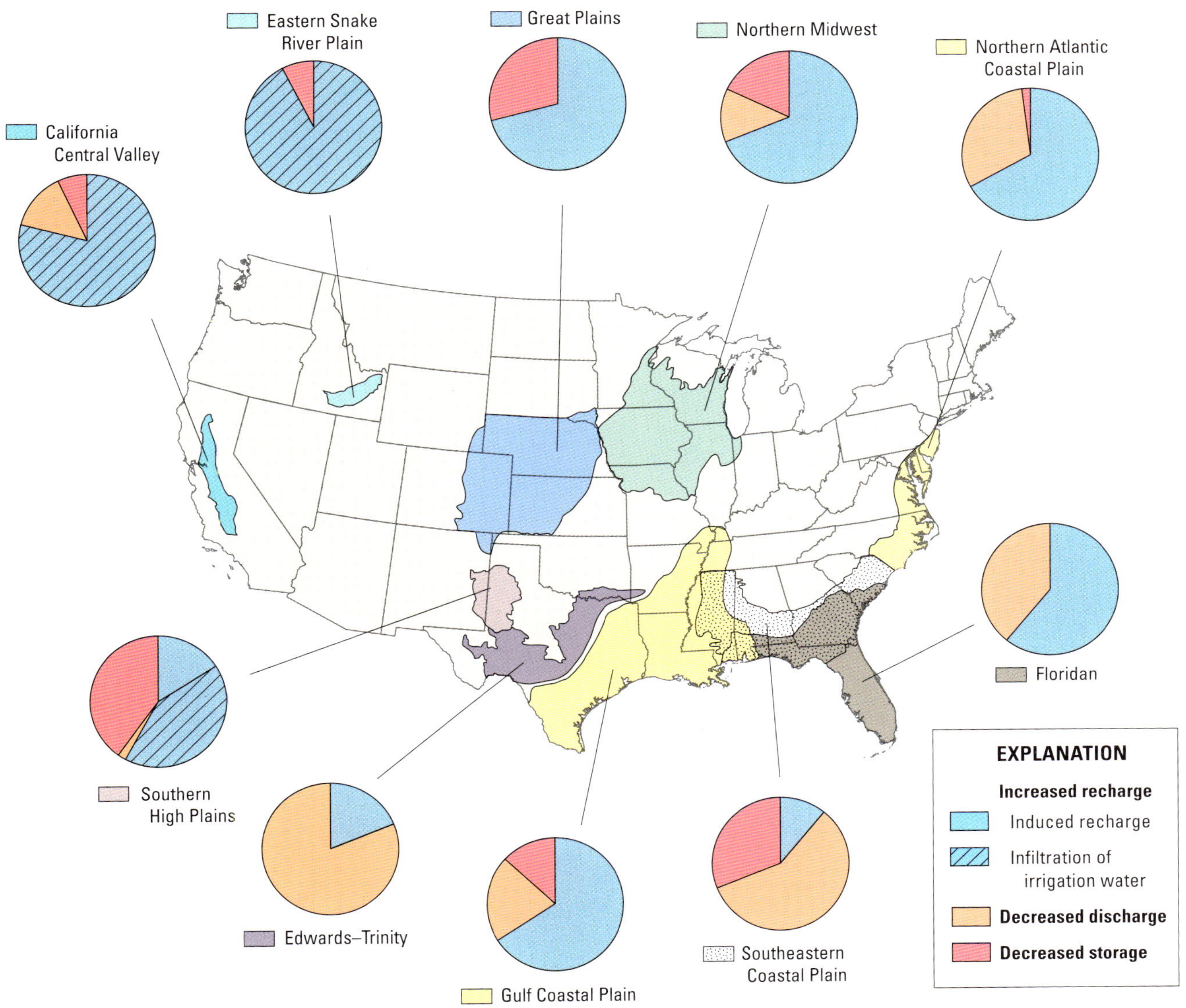

Figure 22. The sources of water that supply withdrawals from selected major aquifer systems in the United States based on model simulations for various periods through the 1970s and 1980s. The results illustrate the variety of ways in which overall ground-water budgets can change in response to large-scale pumping. The simulations of the aquifer systems were for different stages in the development of each system and, in many cases, are not representative of current (2007) conditions given the dynamic nature of ground-water systems (modified from Alley and others, 2002).

The principal aquifers shown in figure 23 were defined on the basis of geologic and hydrologic characteristics. Some of the principal aquifers, such as the California Central Valley aquifer system, define a regional ground-water flow system. Other principal aquifers do not define a single distinct regional flow system but provide a good starting point for defining regional aquifer systems to be analyzed for national assessment. In these cases, the studies undertaken would include combinations of parts of multiple principal aquifers that define a regional flow system. For example, the alluvial aquifers, which make up a principal aquifer distributed throughout the United States, would be part of many of the regional aquifer systems evaluated. Examples of how regional aquifer systems can be defined from the principal aquifers are provided in several case studies described in the section on "Examples of Regional Aquifer Assessments."

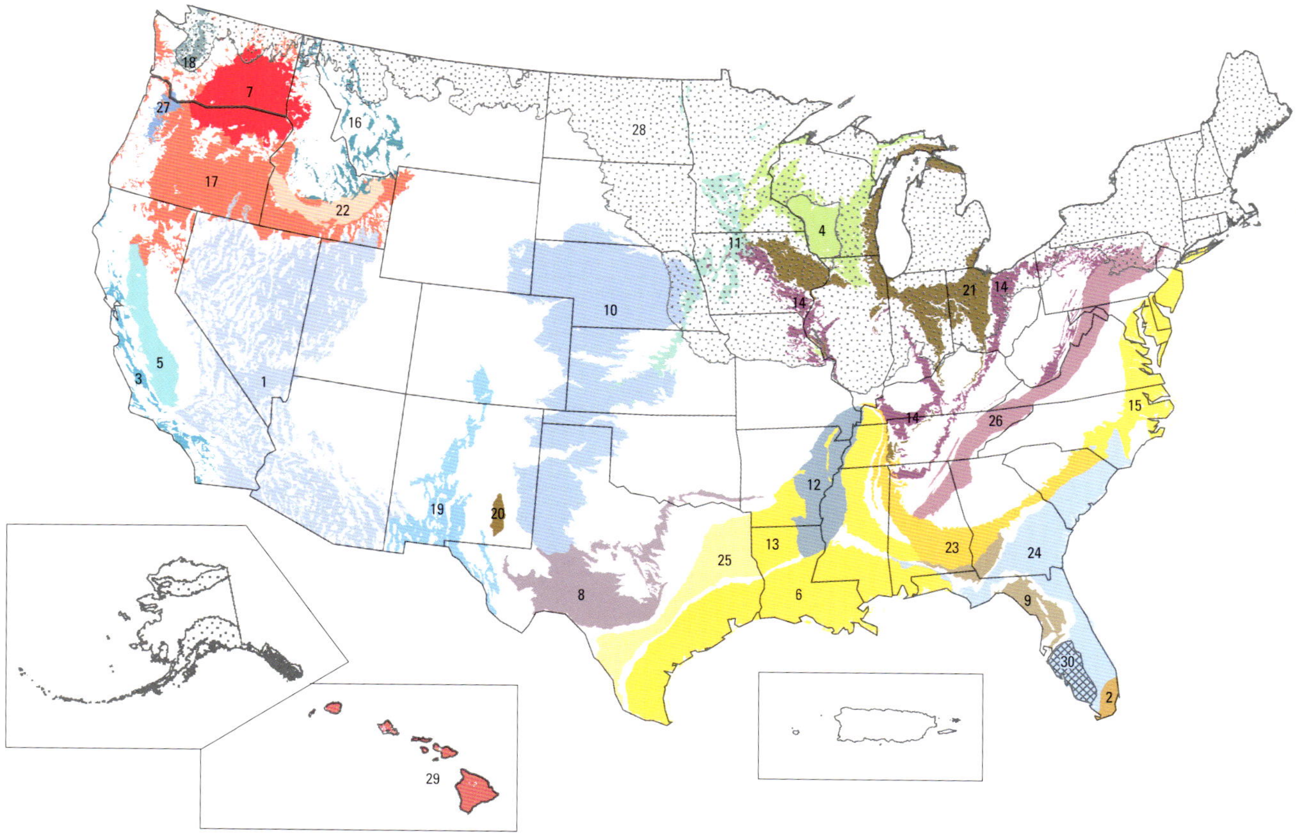

EXPLANATION

Thirty of the principal aquifers

- 1 Basin and Range basin-fill and carbonate-rock aquifers
- 2 Biscayne aquifer
- 3 California Coastal Basin aquifers
- 4 Cambrian–Ordovician aquifer system
- 5 Central Valley aquifer system
- 6 Coastal lowlands aquifer system
- 7 Columbia Plateau basaltic-rock and basin-fill aquifers
- 8 Edwards–Trinity aquifer system
- 9 Floridan aquifer system
- 10 High Plains aquifer
- 11 Lower Cretaceous aquifers (IA, KS, MN, and NE)
- 12 Mississippi River Valley alluvial aquifer
- 13 Mississippi embayment aquifer system
- 14 Mississippian aquifers
- 15 Northern Atlantic Coastal Plain aquifer system
- 16 Northern Rocky Mountains intermontane basins aquifer system
- 17 Pacific Northwest basaltic-rock and basin-fill aquifers
- 18 Puget Sound aquifer system
- 19 Rio Grande aquifer system
- 20 Roswell Basin aquifer system
- 21 Silurian–Devonian aquifers
- 22 Snake River Plain basaltic-rock and basin-fill aquifers
- 23 Southeastern Coastal Plain aquifer system
- 24 Surficial aquifer system
- 25 Texas coastal uplands aquifer system
- 26 Valley and Ridge carbonate-rock and other aquifers
- 27 Willamette Lowland basin-fill aquifers
- 28 Glacial sand and gravel aquifers
- 29 Volcanic-rock aquifers (Hawaii)
- 30 Intermediate aquifer system (Florida)

Figure 23. Thirty principal aquifers that collectively account for about 94 percent of the Nation's total ground-water withdrawals for public supply, irrigation, and self-supplied industrial uses combined.

Table 1. The 30 regional principal aquifers with the greatest amount of ground-water use. Water-use estimates from Maupin and Barber (2005).

Regional principal aquifer	Principal aquifer number in figure 2	Total water use	Irrigation	Public supply	Self-supplied industrial
		Million gallons per day			
High Plains aquifer	11	17,488	17,000	389	99
Central Valley aquifer system	10	9,808	8,910	839	59
Mississippi River Valley alluvial aquifer	13	9,290	9,150	70	70
Basin and Range basin-fill and carbonate-rock aquifers	1 and 42	5,695	4,550	1,080	65
Glacial sand and gravel aquifers	shaded area	4,075	1,170	2,273	632
Floridan aquifer system	50	3,645	1,930	1,330	385
California Coastal Basin aquifers	3	3,446	1,760	1,580	106
Snake River Plain basaltic-rock and basin-fill aquifers	6 and 59	3,075	2,900	151	24
Coastal lowlands aquifer system	18	2,368	933	1,010	425
Pacific Northwest basaltic-rock and basin-fill aquifers	4 and 58	1,340	1,206	121	13
Rio Grande aquifer system	2	1,119	867	240	12
Columbia Plateau basaltic-rock and basin-fill aquifers	5 and 60	1,077	810	223	44
Northern Atlantic Coastal Plain aquifer system	22	1,035	70	793	172
Mississippi embayment aquifer system	20	946	195	576	175
Cambrian–Ordovician aquifer system	33	933	92	590	251
Southeastern Coastal Plain aquifer system	21	860	382	340	138
Biscayne aquifer	51	812	114	698	0
Edwards–Trinity aquifer system	38	740	282	411	47
Surficial aquifer system (southeastern United States)	15	650	364	263	23
Volcanic-rock aquifers (Hawaii)	61	429	171	243	15
Willamette Lowland basin-fill aquifers	8	420	245	99	76
Roswell Basin aquifer system	43	386	364	21	1
Texas coastal uplands aquifer system	19	381	188	148	45
Northern Rocky Mountains inter-montane basins aquifer system	9	377	264	78	35
Valley and Ridge carbonate-rock and other aquifers	39	363	7	226	130
Intermediate aquifer system (Florida)	not shown	354	292	61	1
Lower Cretaceous aquifers (Iowa, Kansas, Minnesota, and Nebraska only)	25	317	259	53	5
Mississippian aquifers	40	285	6	211	68
Puget Sound aquifer system	7	260	45	192	23
Silurian–Devonian aquifers	47	246	27	164	55

Table 2. Previous investigations of the 30 regional principal aquifers with the greatest amount of ground-water use.

[Acronyms used in the table are: RASA, Regional Aquifer-System Analysis Program; NAWQA, National Water-Quality Assessment Program; and GWRP, Ground-Water Resources Program]

Regional principal aquifer	Dates of RASA study that included area	Recent regional investigations	Selected references for previous regional studies
High Plains aquifer	1978–86	Selected areas have had hydrologic investigations. A water-quality investigation under the NAWQA program took place from 1999 to 2006.	McMahon and others, 2007; Luckey and Becker, 1999; Weeks and others, 1988
Central Valley aquifer system	1978–82	A followup geochemical investigation (RASA Phase 2) was completed in 1990. A GWRP study started in 2004 and has recently been completed.	Bertoldi and others, 1991; Claudia C. Faunt, U.S. Geological Survey, written commun., 2007
Mississippi River Valley alluvial aquifer	1980–91	Followup quantitative studies have been conducted in Arkansas.	Ackerman, 1996; Czarnecki, Clark, and Reed, 2003; Czarnecki, Clark, and Stanton, 2003; Grubb, 1998; Reed, 2003; Stanton and Clark, 2003
Basin and Range basin-fill and carbonate-rock aquifers	1978–1990 (Arizona); 1980–88 (Nevada and Utah)	Selected areas have been studied since the RASA Program including Death Valley Regional Flow System, Antelope Valley–Western Mojave, Mojave River Ground-Water Basin, Joshua Tree Ground-Water Subbasin–Mojave Desert, Northern Utah Valley, and Upper San Pedro Basin, Arizona.	Anderson, 1995; Belcher, 2004; D'Agnese and others, 2002; Harrill and Prudic, 1998; Leighton and Phillips, 2003; Nishikawa and others, 2005; Pool and Dickinson, 2007; Stamos and others, 2001; Thiros, 2006
Glacial sand and gravel aquifers	1981–89	The RASA study covered the Northeast Glacial aquifer system. The remainder of the glacial system has not had a systematic flow analysis but is the subject of a current water-quality investigation under the NAWQA program. Selected areas that were not covered by the RASA study have been studied, including Colville Basin, Washington; Spokane Valley–Rathdrum Prairie, Washington–Idaho, and Methow Basin Washington. A current study of water availability in the Great Lakes basin will include the glacial aquifer system.	Ely and Kahle, 2004; Grannemann and Reeves, 2005; Hsieh and others, 2007; Konrad and others, 2005; Randall, 2001; Warner and Arnold, 2005
Floridan aquifer system	1978–86	Several recent regional investigations have been carried out in Georgia and Florida since the RASA study, however, they have not covered the entire aquifer.	Johnston and Bush, 1988; Payne and others, 2005, 2006; Provost and others, 2006; Sepulveda, 2002
California Coastal Basin aquifers	1989–94	A RASA study included intensive study of the Santa Clara–Calleguas Basin. Since then, substantial work has been done on other coastal basins including Pajaro, Santa Barbara, Los Angeles, San Bernardino, and Santa Clara Valley.	Danskin and others, 2006; Freckleton and others, 1998; Hanson, 2003; Hanson and others, 2003; Hanson and others, 2004; Reichard and others, 2003
Snake River Plain basaltic-rock and basin-fill aquifers	1979–90		Lindholm, 1996
Coastal lowlands aquifer system	1980–91	A followup study, including a subsidence model, was conducted for the northern part of the aquifer system in Texas.	Grubb, 1998; Kasmarek and Robinson, 2004; Martin and Whiteman, 1999
Pacific Northwest basaltic-rock and basin-fill aquifers		Studies of the Upper Deschutes Basin and the Portland Basin have been undertaken.	Gannett and Lite, 2004; Morgan and McFarland, 1996
Rio Grande aquifer system	1978–90	A comprehensive study of ground water in the Middle Rio Grande Basin, which is the basin that includes Albuquerque, New Mexico, was undertaken from 1995 to 2000.	Bartolino and Cole, 2002; Wilkins, 1998
Columbia Plateau basaltic-rock and basin-fill aquifers	1982–89		Vaccaro, 1999
Northern Atlantic Coastal Plain aquifer system	1979-87	Since the RASA study, selected areas have had hydrologic investigations. The GWRP study of the North Carolina–South Carolina area has recently been completed in 2007. In New Jersey, the RASA model has been updated and numerous smaller studies have been conducted. A study of the Coastal Plain Aquifer in the Mid-Atlantic area is in progress.	Ator and others, 2005; Bruce G. Campbell, U.S. Geological Survey, written commun., 2007; Shedlock and Bolton, 2006; Trapp and Meisler, 1992; Voronin, 2004

Table 2. Previous investigations of the 30 regional principal aquifers with the greatest amount of ground-water use.—Continued

[Acronyms used in the table are: RASA, Regional Aquifer-System Analysis Program; NAWQA, National Water-Quality Assessment Program; and GWRP, Ground-Water Resources Program]

Regional principal aquifer	Dates of RASA study that included area	Recent regional investigations	Selected references for previous regional studies
Mississippi embayment aquifer system	1980–91	Followup studies and models of the Sparta aquifer have been conducted in southeast Arkansas and north-central Louisiana.	Arthur and Taylor, 1998; Grubb, 1998; McKee and Clark, 2003; McKee and others, 2004
Cambrian–Ordovician aquifer system	1978–84	Since the RASA study, selected areas have been studied. In Wisconsin, a 10-county regional ground-water assessment was conducted. The current study of water availability in the Great Lakes basin will include the Cambrian–Ordovician aquifers.	Feinstein and others, 2005; Grannemann and Reeves, 2005; Young, 1992
Southeastern Coastal Plain aquifer system	1979–88	Since the RASA study, selected areas have had hydrologic investigations. The GWRP study of the North Carolina–South Carolina area was recently completed in 2007.	Bruce G. Campbell, U.S. Geological Survey, written commun., 2007; Miller, 1992
Biscayne aquifer		A numerical model was completed in 2001.	Langevin, 2001
Edwards–Trinity aquifer system	1986–94	Since the RASA study, significant work has been undertaken studying the Edwards aquifer and a three-dimensional ground-water model was completed in 2004.	Lindgren and others, 2004
Surficial aquifer system (southeastern United States)			
Volcanic-rock aquifers (Hawaii)	1982–88	The RASA study was of the island of Oahu. A simulation of the aquifer in the Pearl Harbor area was completed in 2005. Studies on other islands have been done since the RASA Program, including Molokai; Kona, Hawaii; Hawi aquifer, Hawaii; and Southern Lihue Basin, Kauai.	Izuka and Gingerich, 1998; Izuka and Oki, 2002; Nichols and others, 1997; Oki, 1997, 1999, 2002, 2005, 2006; Underwood and others, 1995
Willamette Lowland basin-fill aquifers	1989–94	There has been a recent geologic and hydrogeologic framework study.	Conlon and others, 2005; Gannett and Caldwell, 1998
Roswell Basin aquifer system			
Texas coastal uplands aquifer system	1980–91		Grubb, 1998
Northern Rocky Mountains intermontane basins aquifer system		A RASA study was planned and begun in 1992 but not completed because of budget cuts.	Clark and Kendy, 1992
Valley and Ridge carbonate-rock and other aquifers	1988–93	The RASA study of the Appalachian Piedmont, Valley and Ridge, and Blue Ridge Provinces included simulation of the Cumberland Valley in Pennsylvania (PA). Several local-scale studies have been conducted in PA, Maryland, Tennesssee, West Virginia (WV), and Virginia (VA). A study of the Shennandoah Valley in VA and WV is in progress.	Chichester, 1996; Haugh, 2002; Swain and others, 2004
Intermediate aquifer system (Florida)		A study was recently published in 2006.	Knochenmus, 2006
Lower Cretaceous aquifers (IA, KS, MN and NE only)		State-wide studies have been done on different parts of the system (IA, Iowa; KS, Kansas; MN, Minnesota; NE, Nebraska).	Burkart 1984; Woodward and Anderson, 1986
Mississippian aquifers			
Puget Sound aquifer system	1989–94	The hydrogeologic framework has been studied and there has been a model analysis of Thurston County.	Drost and others, 1999; Vaccaro and others, 1998
Silurian–Devonian aquifers	1988–91	The RASA study of the Midwestern Basins and arches glacial and carbonate regional aquifer system included part of the Silurian–Devonian aquifers. Two studies of the Silurian–Devonian aquifer in a part of southeastern Michigan and in part of Iowa were recently completed. A current study of water availability in the Great Lakes basin will include the Silurian–Devonian aquifers.	Bugliosi, 1999; Grannemann and Reeves, 2005; Reeves and others, 2004; Tucci and McKay, 2006

C

The U.S. Geological Survey Regional Aquifer-System Analysis (RASA) Program

The Regional Aquifer-System Analysis (RASA) Program began in response to the 1977 drought and recommendations by the U.S. National Water Commission and the U.S. Comptroller General. From 1978 to 1995, 25 of the Nation's most important ground-water systems were evaluated as part of the RASA Program (fig. C1). Computer models were used to develop estimates of current and future water availability for many of these systems. In addition, the Ground-Water Atlas of the United States (Miller, 2000; *http://capp.water.usgs.gov/gwa/*) was compiled as a general source of information on ground-water resources.

The RASA Program provided a baseline of knowledge on the aquifer systems studied that will prove useful for many decades to come. However, as valuable as the RASA Program was, it leaves some important needs unfulfilled. The program did not examine many of the shallower or less productive aquifers that are very important to rural and small community water users and in sustaining flow in streams. The RASA Program also was static, describing the aquifers at a point in time. Virtually all of the information used in the RASA ground-water studies is now more than 15 to 20 years old. Ground water is dynamic, and aquifers need to be re-examined over time as conditions and issues change.

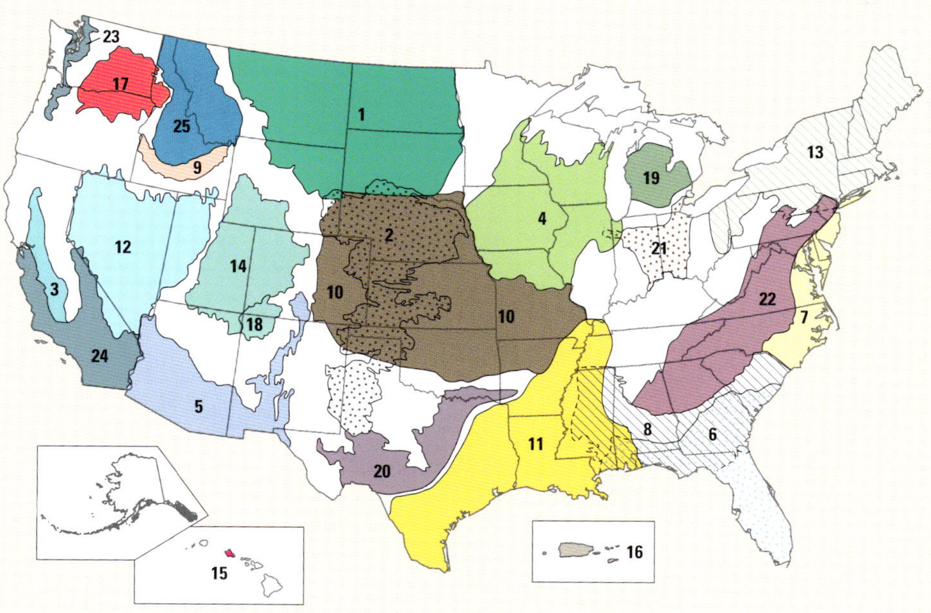

EXPLANATION

RASA study areas

1 Northern Great Plains
2 High Plains
3 Central Valley, California
4 Northern Midwest
5 Southwest alluvial basins
6 Floridan
7 Northern Atlantic Coastal Plain
8 Southeastern Coastal Plain
9 Snake River Plain
10 Central Midwest
11 Gulf Coastal Plain
12 Great Basin
13 Northeast glacial aquifers
14 Upper Colorado River Basin
15 Oahu, Hawaii
16 Caribbean Islands
17 Columbia Plateau
18 San Juan Basin
19 Michigan Basin
20 Edwards–Trinity
21 Midwestern basins and arches
22 Appalachian valleys and Piedmont
23 Puget–Willamette Lowland
24 Southern California alluvial basins
25 Northern Rocky Mountain Intermontane Basins

Figure C1. The 25 regional aquifers studied under the Regional Aquifer-System Analysis (RASA) Program from 1978 to 1995 (from U.S. Geological Survey, 1998).

Regional Studies

The ground-water assessment undertaken for each regional ground-water flow system emphasizes the integrated use of monitoring data, ground-water modeling, and other existing information to place the status of ground-water resources in the context of the complete water budget for that aquifer system (see Box D). This assessment is structured to provide a national perspective on the Nation's ground-water resources, while simultaneously emphasizing the value of the regional assessments to those who manage and use the resources regionally and locally. The previous RASA assessments are mostly more than 20 years old and the new proposed assessments would build on the information developed under the RASA Program where appropriate but take advantage of new information and techniques to assess the ground-water resources of the Nation.

As a first step in the investigations, existing information would be compiled and synthesized on ground-water availability for the entire regional ground-water flow system. This information includes what is known about changes in ground-water levels, storage, recharge, and discharge and identification of major regional ground-water availability issues. The second step would involve regional ground-water modeling to estimate historic changes in water budgets and provide a tool for estimating system response to future stresses. Depending on the situation, the entire regional aquifer system or selected parts of the regional aquifer system would be modeled. For example, in some cases, a regional aquifer system may consist of many individual basins complicating meaningful simulation of the entire aquifer system. In other cases, information is most needed on particular parts of the aquifer system. If previous ground-water flow models exist for the area or part of the area, they could be used as a starting point to develop more up-to-date and accurate models. More focused studies would be used to understand processes and effects (for example, stream-aquifer interactions) that are important regionally but occur at finer scales. Regional modeling studies would be emphasized to the extent practicable, so that the results from all regional aquifer studies could be assembled to provide a national assessment. Development and testing of new approaches to regional assessment would be an important part of the studies to advance the approaches taken over time.

As noted in Box D, the completion of model analysis of aquifer systems provides an opportune time to summarize insights on the value of the existing monitoring networks and possible gaps in coverage. Thus, the regional modeling studies would provide feedback to others on existing monitoring networks.

Overall, products from the regional studies would include:

1. Water budgets of major aquifer systems;
2. Current estimates and historic trends in ground-water use, storage, recharge, and discharge;
3. Ground-water models that provide a regional context for more local studies and a tool for others to make future projections of ground-water availability;
4. Regionwide estimates of key hydrologic variables (for example, aquifer properties and recharge) for major aquifers;
5. An evaluation of the existing networks for monitoring ground-water availability; and
6. Testing and evaluation of new approaches for analysis of regional aquifers.

These products would provide the foundational information and modeling tools to help State and local agencies make water-availability decisions based on local water-management constraints and goals.

The regional studies would build on a foundation of previous and ongoing ground-water studies. Additionally, whenever possible, future ground-water availability studies would be scheduled to take advantage of other coincident efforts for the purpose of leveraging resources to the maximum extent possible. Each regional study would require 3 to 6 years for completion, depending on complexity. Given the current funding level of the USGS Ground-Water Resources Program, a national assessment would require 2 to 3 decades, although interim information would likely be released periodically to summarize the state of knowledge at that time.

Water availability involves both the quantity and quality of the resource. Interpretation of ground-water quality is best achieved with an understanding of ground-water flow systems. Thus, various ways would be explored to link studies of the NAWQA Program with the hydrologic studies of the regional aquifer systems. Likewise, improved understanding

D

Relation of Ground-Water Modeling to Assessment and Monitoring

A ground-water model attempts to reproduce, or simulate, the operation of a ground-water system with a mathematical counterpart. Models are commonly used to evaluate changes to the water budget of an aquifer resulting from land-use changes, water withdrawals, and climate, and how these changes affect streamflow, lake levels, water quality, and other important variables.

An important role of ground-water modeling is to place current conditions defined by monitoring data in the context of the slow changes that may be taking place in the hydrologic system. Many aquifer systems have undergone several decades of development and may be far from equilibrium. Data on current conditions may not indicate, for example, how future streamflow depletion will evolve from the pumping that has already occurred, but this can be estimated by the use of models.

Monitoring and computer modeling are complementary activities, but too often are treated separately, ignoring important linkages and feedbacks. An idealized framework for integration of monitoring and modeling in the context of ground-water assessment is illustrated in figure D1. In this figure, the top row signifies a long-term network that is systematically monitored over time. The second row represents the development of models and their periodic updates to advance understanding of how the aquifer system responds to human development, integrate new information from scientific studies, and address new questions as they may arise. Monitoring data serve as primary information for calibration of computer models. Conversely the process of model calibration and use provides insights into the adequacy of and gaps in monitoring data. This is shown by the arrows representing long-term monitoring as input to modeling and a feedback loop to evaluate long-term monitoring networks on the basis of modeling. Unfortunately, the second step, evaluation of monitoring networks at the conclusion of a modeling study too rarely occurs.

Figure D1 explicitly recognizes that every simulation model is built upon an underlying conceptual model of how the ground-water system works. More often than not, data will fit more than one conceptual model, and good calibration of a model does not ensure a correct conceptual model (Bredehoeft, 2003). The appropriateness of the conceptual model is tested as a ground-water model is built, and field observations are compared to the model simulations. The conceptual model of a system should be evaluated periodically and updated as an important part of updating models.

The final row in figure D1 signifies that periodic studies in addition to long-term monitoring networks should be integrated into each stage of model development. For example, information about water sources and the age of the water (time since recharge) obtained from environmental tracers can be compared to ground-water ages and flow paths inferred from modeling. Likewise, geologic and geophysical studies may provide new insights into the hydrogeologic framework.

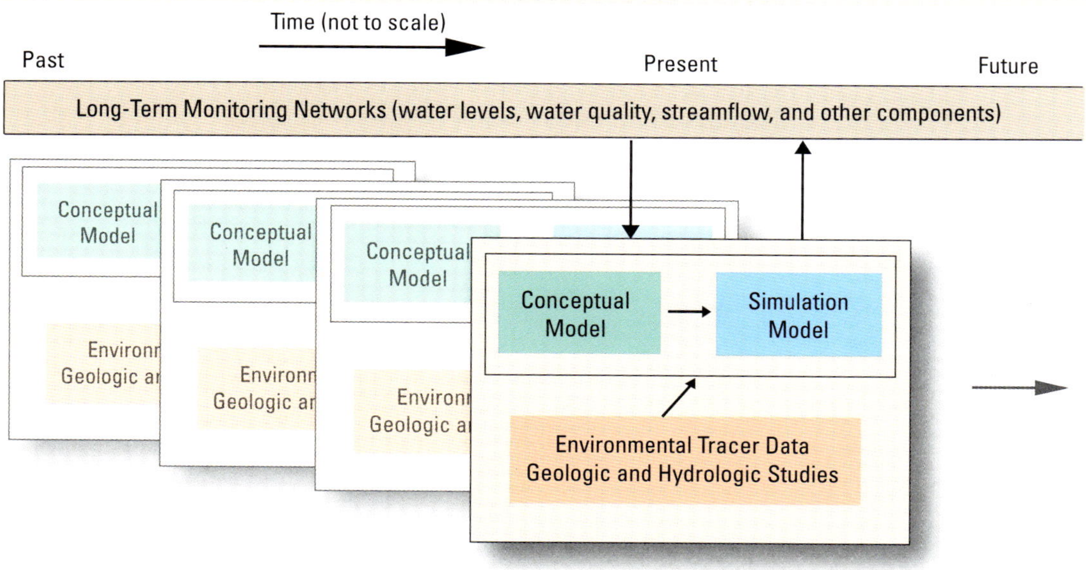

Figure D1. A framework for integration of monitoring and modeling in the context of ground-water assessment (modified from Alley, 2006).

of geologic frameworks provided by the USGS National Cooperative Geologic Mapping Program and other geologic mapping by the State Geological Surveys would be important input to the regional-scale models developed as part of the national assessment. Complementary data collection and sharing of information among the USGS and State and local agencies would enhance the ability of the regional studies to be relevant and useful to those managing the resources.

Examples of Regional Aquifer Assessments

Five case studies (fig. 24) are used to illustrate the diversity of water-availability issues, the approaches to regional assessment, and the resultant information produced by these studies. All but one of the case studies include overall water budgets for the aquifer system under predevelopment and development conditions.

The first case study summarizes a comprehensive multiagency evaluation of the Middle Rio Grande Basin undertaken in the late 1990s. The next two case studies (California Central Valley and Coastal Plain) are recently completed regional aquifer assessments by the USGS Ground-Water Resources Program. The fourth case study features an ongoing study in the Great Lakes Basin, with a focus on the Lake Michigan ground-water basin. Finally, the High Plains aquifer, although lacking a recent ground-water model for the complete system, is used to illustrate the value of a long-term water-level monitoring program and the relation of water quality to water availability at the regional scale.

Rose M. McGowan, USGS

Lake Michigan from Manistique, Michigan

Claudia C. Faunt, USGS

An orchard in California's Central Valley

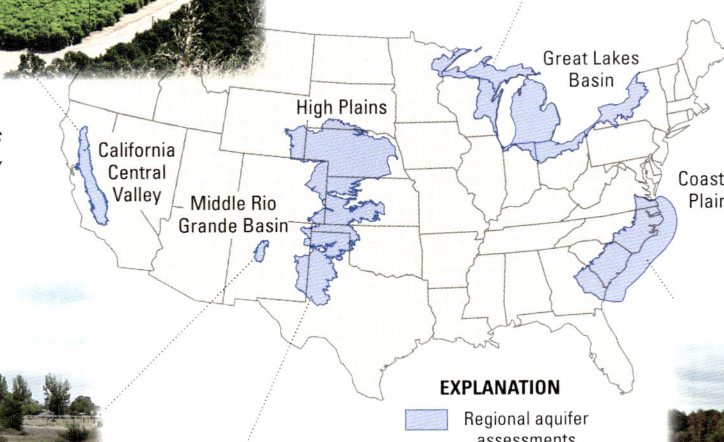

Figure 24. Five case studies that document the assessment of regional ground-water systems.

James R. Bartolino, USGS

The Bernalillo Riverside Drain, New Mexico

Richard R. Luckey, USGS

Center-Pivot Irrigation on the High Plains

Thomas A. Abrahamsen, USGS

May River in Beaufort County, South Carolina

Middle Rio Grande Basin

Introduction

The Middle Rio Grande Basin is an area of about 3,060 mi^2 in central New Mexico within the Rio Grande valley extending from about Cochiti Lake downstream to about San Acacia (fig. 25). The Middle Rio Grande Basin is part of the principal aquifer referred to as the "Rio Grande aquifer system" in figure 23. The Middle Rio Grande Basin covers about 10 percent of this principal aquifer and includes an area of considerable ground-water use by the City of Albuquerque and surrounding communities. The climate over most of the basin is semiarid, with mean annual precipitation ranging from 7.6 inches at Belen to 12.7 inches at Cochiti Dam (Bartolino and Cole, 2002).

Most water-bearing units of the Middle Rio Grande Basin are unconsolidated deposits of the Tertiary-age Santa Fe Group. The Tertiary-age Santa Fe Group deposits are basin-fill deposits representing the transition from a topographically closed basin to the current through-flowing Rio Grande. Because the Santa Fe Group and post-Santa Fe Group deposits are hydraulically connected, they are commonly grouped together as the Santa Fe Group aquifer system (fig. 26).

The thickness of the Santa Fe Group in the Middle Rio Grande Basin is highly variable because of complex faulting during sedimentation. Total thickness ranges from about 1,400 ft at basin margins to approximately 14,000 ft in localized areas in the center of the basin (Bartolino and others, 2002). Ground water is withdrawn mostly from the sands and gravels of the upper and middle parts of the aquifer; only about the upper 2,000 ft of the aquifer is used for ground-water withdrawal. The depth to water in the aquifer system varies widely from about 0 ft near the Rio Grande to as much as 1,180 ft in an area west of Albuquerque.

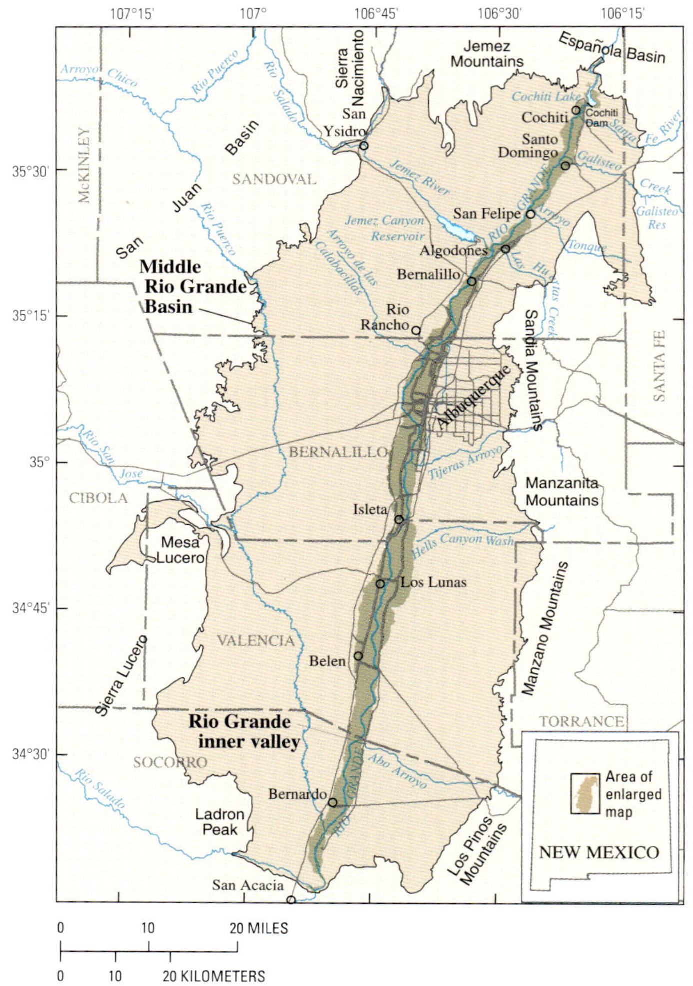

Figure 25. Major physiographic and hydrologic features of the Middle Rio Grande Basin, New Mexico (modified from Bartolino and others, 2002).

Water Availability Issues

In 2000, the population of the Middle Rio Grande Basin was about 690,000 or about 38 percent of the population of New Mexico. Water for municipal and domestic supply is currently (2007) almost exclusively from ground water. The New Mexico Office of the State Engineer administers the appropriation and use of the water resources of New Mexico and has declared the basin a "critical basin;" that is, a ground-water basin faced with rapid economic and population growth where there is less than adequate technical information as to the available water supply (Bartolino and others, 2002). In addition, surface-water flows of the Rio Grande are considered fully appropriated by the New Mexico Office of the State Engineer, and an equivalent surface-water right must be obtained to offset any ground-water withdrawals that deplete the river (Bartolino and Cole, 2002, p. 69). In this semiarid environment, understanding the amount of water available for use and the interaction between surface water and ground water is critical to planning for the future.

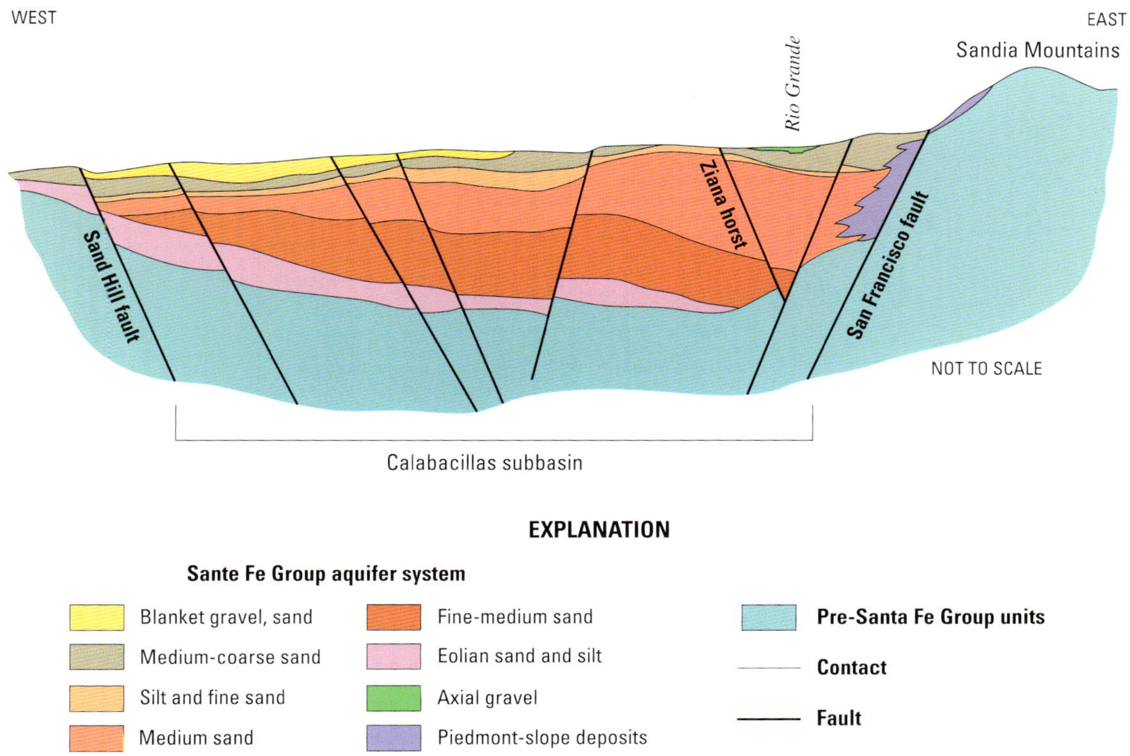

Figure 26. Generalized west-east cross section of hydrogeologic deposits in the Middle Rio Grande Basin (from Bartolino and Cole, 2002).

Approach to Assessment

The Middle Rio Grande Basin study was a 6-year effort by the USGS and other agencies to improve the understanding of the hydrology, geology, and land-surface characteristics of the Middle Rio Grande Basin (Bartolino and Cole, 2002). The approach undertaken was to develop an up-to-date understanding of the hydrogeology, ground-water quality, water use, mountain recharge, and hydraulic connection between the Rio Grande and the ground-water system.

The conceptual geologic framework of the Middle Rio Grande Basin was revised and updated by mapping the surficial and bedrock deposits of the Middle Rio Grande Basin and adjoining areas. Several different geophysical methods were used to aid in developing the revised hydrogeologic framework. In particular, high-resolution aeromagnetic surveys delineated faults that offset water-bearing units in the aquifer system, which may play an important role in the ground-water flow system (fig. 27). Ground-water samples were extensively analyzed for environmental tracers and other chemical constituents in order to date ground water, to define zones of differing water quality, and to locate areas of recent recharge (Plummer and others, 2004). The interaction between the Rio Grande and the ground-water system was estimated by electromagnetic surveys and measurement of streamflow losses. Research on mountain front recharge indicated that the amount of water being recharged to the ground-water system is less than was previously estimated (Anderholm, 2000).

All of these improvements in the understanding of the ground-water system were incorporated in a quantitative ground-water flow model of the basin (McAda and Barroll, 2002). The model enables estimates to be made of the water budget of the ground-water system under predevelopment conditions and through time as human influences changed the system.

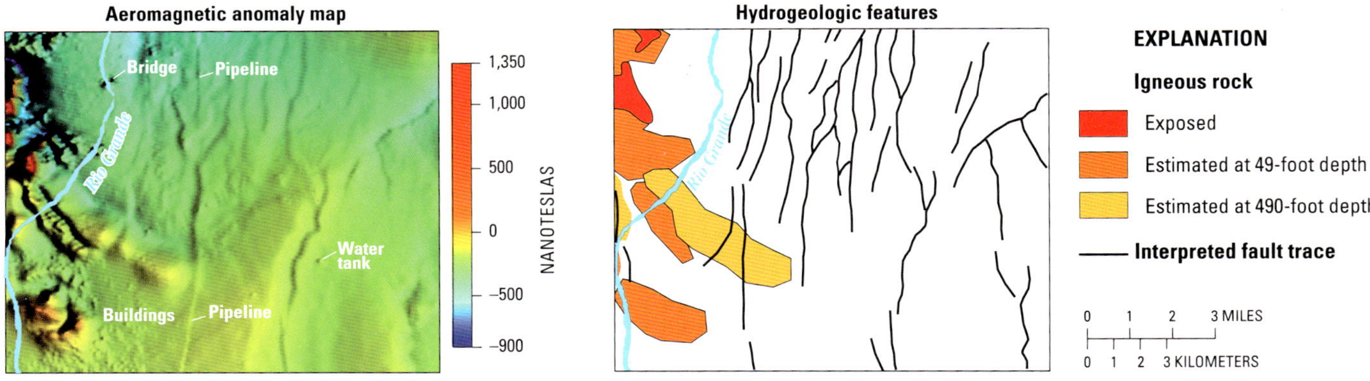

Modified from U.S. Geological Survey and Sander Geophysics, Ltd., 1998

Figure 27. Aeromagnetic anomaly map of an area south of Albuquerque and simplified map of important hydrogeologic features. Many geologic features and human-made structures can be seen on the anomaly map, which is displayed in color and shaded as though it were a relief map illuminated from the east. The most evident hydrogeologic features on the aeromagnetic map are faults and igneous rocks, which are depicted on the simplified map (from Bartolino and others, 2002).

Selected Findings

Before major human influences (predevelopment), the ground-water system in the Middle Rio Grande Basin was in an approximate state of dynamic equilibrium. All the water recharging the system naturally discharged from the system. As estimated by the ground-water flow model, predevelopment recharge to the aquifer from the Rio Grande, Jemez River, other rivers, mountain front recharge, and underflow from neighboring basins amounted to approximately 200 cubic feet per second (ft^3/s) (McAda and Barroll, 2002; Douglas P. McAda, U.S. Geological Survey, written commun., 2007), and most of the water discharged as evapotranspiration (fig. 28). Once humans began development of the basin, canals were built for irrigation, wells were drilled for supply, drains were built to prevent water logging of soils, and water was removed from storage. These changes drastically altered the hydrologic budget in the ground-water system.

Ground-water levels have declined in many parts of the Middle Rio Grande Basin; the water table has declined more than 160 ft since 1945 in some areas (Bartolino and Cole, 2002). Ground-water flow model estimates (McAda and Barroll, 2002; Douglas P. McAda, U.S. Geological Survey, written commun., 2007) for 1999 indicate that total inflows (including water from storage) and total outflows increased to about 560 ft^3/s, with withdrawals from wells accounting for more than 200 ft^3/s of the outflows. The flow rates presented in McAda and Barroll (2002) for the Rio Grande and Jemez River were net flow rates that combined inflows and outflows; additional information (Douglas P. McAda, U.S. Geological Survey, written commun., 2007) allowed for the separation of inflows and outflows as shown in figure 28. Recharge to the aquifer system is from the runoff from the mountain fronts, intermittent tributaries, underflow from bounding ground-water systems, canal seepage, the Rio Grande, and the

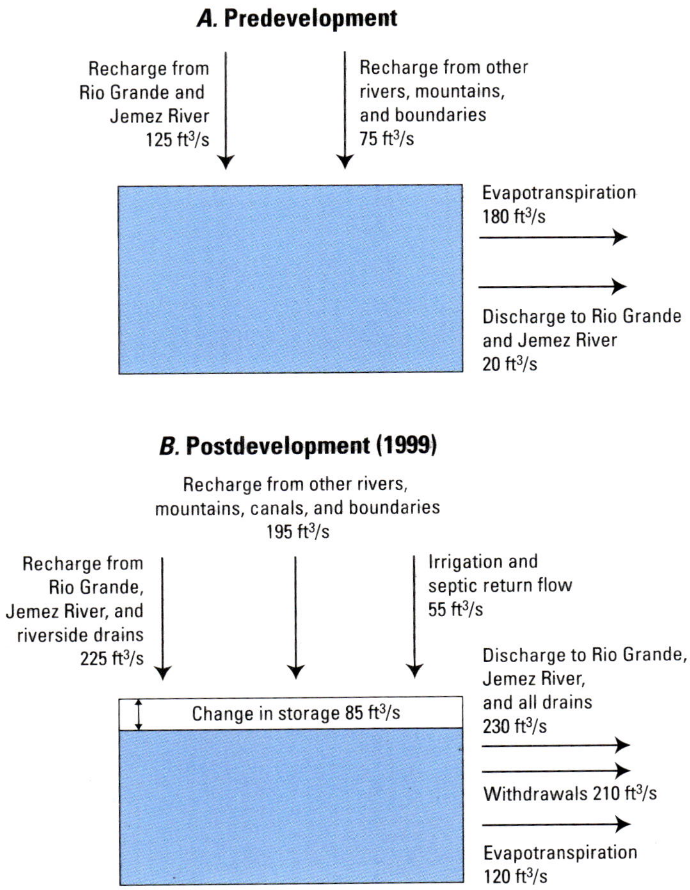

Figure 28. Ground-water budgets before development and postdevelopment (annual rates during 1999) for the Middle Rio Grande Basin aquifer system: *(A)* predevelopment, and *(B)* postdevelopment (1999) (data from McAda and Barroll, 2002; Douglas P. McAda, U.S. Geological Survey, written commun., 2007). [ft^3/s, cubic feet per second]

Jemez River. As determined from the model, recharge to the ground-water system from the Rio Grande and Jemez River was about 125 ft^3/s during predevelopment conditions and recharge from the Rio Grande, Jemez River, and the constructed riverside drains to the ground-water system became 225 ft^3/s in 1999, and discharge as evapotranspiration decreased from 180 ft^3/s to about 120 ft^3/s. Because of the scale of the model, the Rio Grande and the riverside drains were both represented in the same model cells. Thus, they tend to act as one combined component in the model (Douglas P. McAda, U.S. Geological Survey, written commun., 2007). Interior drains installed to prevent water logging as a result of irrigation became a major discharge mechanism of the system, accounting for about 185 ft^3/s of the 230 ft^3/s of outflow to the Rio Grande, Jemez River, and all drains. The results of the study demonstrate that the ground-water system has changed substantially since predevelopment, and the Rio Grande is a major source of water to the ground-water system in response to development. The model can be used to estimate and understand the consequences of changes in water use on the ground-water system as well as on the exchange of flow between the ground-water system and the Rio Grande.

The Bernalillo Riverside Drain, New Mexico. Photograph by James R. Bartolino, USGS.

California Central Valley Aquifer System

Introduction

For more than 50 years, California's Central Valley has been one of the most productive agricultural regions of the world, which is due in large part to an ample supply of water for irrigation. On less than 1 percent of the total farmland in the United States, the Central Valley supplies 8 percent of the national agricultural output (by value). In an area of about 20,000 mi^2, the Central Valley produces 250 different crops (Great Valley Center, 1998) with an estimated value of 17 billion dollars per year in 2002 (Great Valley Center, 2005) (fig. 29). This irrigated agriculture relies heavily on a combination of water supply from surface-water diversions and ground-water pumping (Bertoldi and others, 1991). Approximately one-sixth of the Nation's irrigated land is in the Central Valley (Bureau of Reclamation, 1994), and about one-eighth of the Nation's ground-water pumpage is from its aquifers (Maupin and Barber, 2005).

The Central Valley is a large structural trough filled with continental sediments that contains most of the Valley's freshwater resources (Berkstresser, 1973; Page, 1973). The average thickness of the continental sediments is about 2,400 ft. Sediments vary in thickness but generally thicken from south to north. Aquifer sediments contain mostly fluvial and interbedded lacustrine deposits. These continental deposits consist predominantly of lenses of gravel, sand, silt, and clay, with more than half of the total thickness composed of fine-grained sediments (Page, 1986). Most of these fine-grained lenses are not areally extensive; however, there is one major unit, the Corcoran Clay member, in the San Joaquin Valley that is worthy of mention because it can locally function as a confining unit (fig. 29). In the past 50 years, the drilling of numerous long-screened irrigation wells through the unit has minimized the influence of the Corcoran Clay on hydraulic heads above and below the clay. Therefore, the Central Valley aquifer can be thought of as a single heterogeneous aquifer in which the vertical movement of water is dependent on the properties of fine-grained sediments and local influences of high-capacity ground-water wells.

Climate in the Central Valley is arid to semiarid with precipitation being greater in the northern part of the valley (13 to 26 inches) than in the southern part (5 to 16 inches). In contrast to low precipitation in the valley, mean annual precipitation in high altitude mountains that nearly surround the valley can be three times greater than what falls on the valley floor. Streamflow is almost entirely dependent on precipitation in the Sierra Nevada and part of the Klamath Mountains to the north (fig. 29). Depth to water is highly variable but typically can be within 100 ft of land surface and sometimes even deeper, particularly in the south. Prior to ground-water development, the direction of flow was toward the center of the valley, and discharge was primarily as evapotranspiration from wetlands or a small amount as seepage to streams. Precipitation was the principal source of recharge to the aquifer. These historical patterns of ground-water discharge and recharge have been significantly altered by agricultural water development. For additional information on the geologic framework and hydrologic setting of the Central Valley, see Bertoldi and others (1991).

Water Availability Issues

In addition to demands on the water system from agriculture, population growth is also placing demands on the water resources. Between 1990 and 2002, about 4 percent of the Central Valley's irrigated farmland was converted to other uses, primarily for housing and other urban uses (Great Valley Center, 2005). In the past 20 years, population in the Central Valley has nearly doubled, and future growth is projected to continue. These large increases in population have intensified the competition for water within the Central Valley and statewide. Other water issues, such as conservation of agricultural lands, conjunctive use, artificial recharge, hydrologic implications of land-use change, and climate change, have added to the complexity of how to evaluate the individual and joint effects of these different factors on overall water availability.

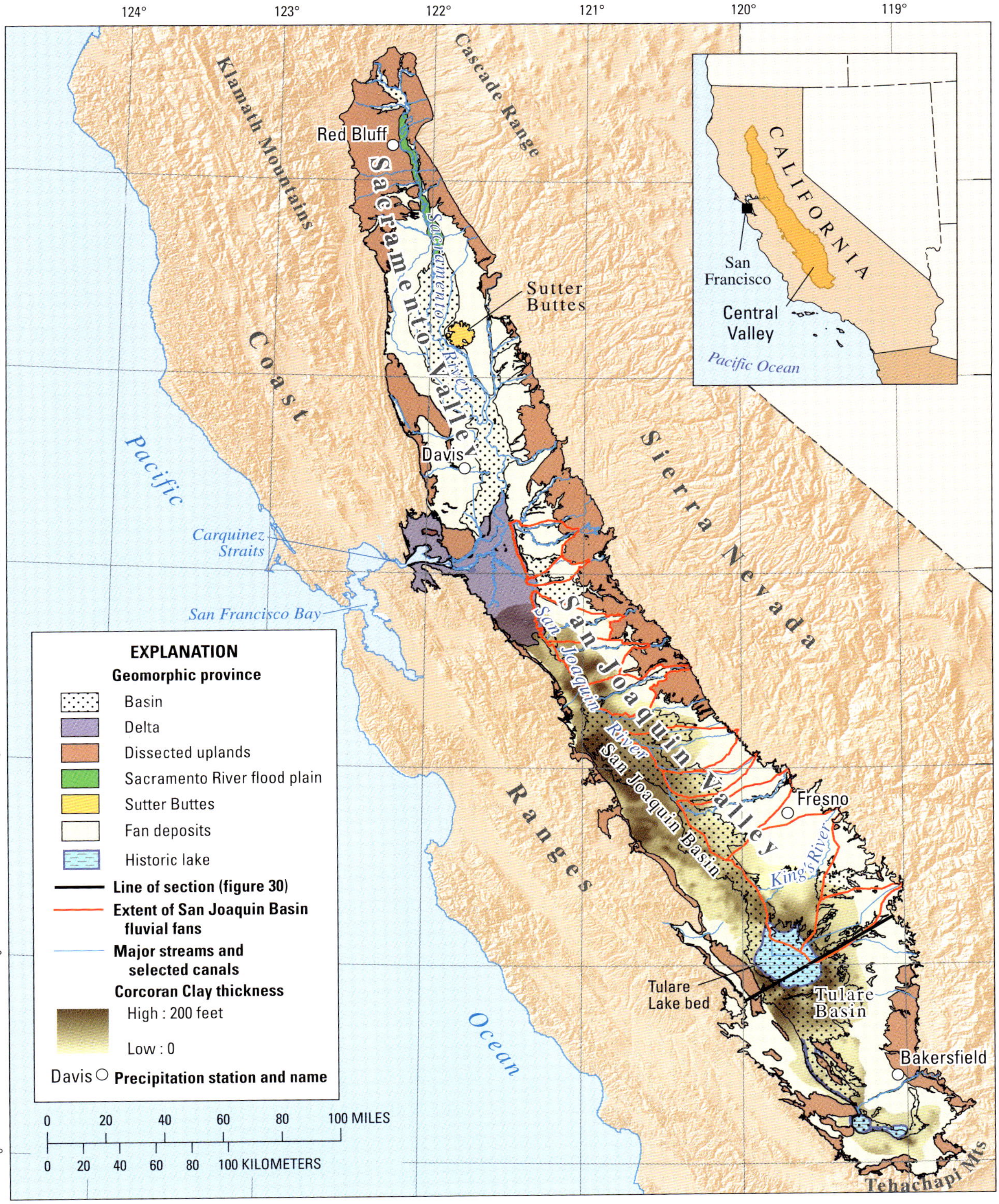

Figure 29. Central Valley aquifer and major geomorphic provinces (modified from Davis and others, 1959; Olmstead and Davis, 1961; and Jennings, 1977), fluvial fans of the San Joaquin Basin (Weissmann and others, 2005), and extent and thickness of Corcoran Clay, California (modified from Page, 1986; Burow and others, 2004).

Approach to Assessment

The Central Valley ground-water system has been studied to varying degrees since about the late 1880s. Starting about the 1940s and continuing for the next four decades, the use of ground water for irrigation increased steadily to a point where ground water was providing about 50 percent of water use for irrigation (Williamson and others, 1989; Bertoldi and others, 1991). This growing dependence on ground water and decrease in the availability of imported surface water because of drought resulted in recognition of the need to begin to investigate ground-water resources across large areas of the valley. The information acquired made it possible to better define geologic features and begin to delineate the hydraulic properties of aquifer sediments. Quantitative assessments using ground-water flow models for parts of the Central Valley aquifer began in the 1970s. However, the first regionally comprehensive model of ground-water flow for the entire Central Valley was conducted by Williamson and others (1989), as part of the USGS RASA Program. The RASA Program provided a wealth of information on the geology, hydrology, and water chemistry of the Central Valley. The ground-water flow model by Williamson and others (1989) simulated conditions from 1961 to 1977, a period of large and variable stresses on the ground-water system. The resulting model provides an overall representation for this large region but is generally inadequate at scales less than about 500 mi^2; a scale at which water-management decisions typically are made.

A variety of investigators from local organizations, State agencies, and private groups have continued to collect data, perform studies, and develop ground-water flow models of various parts of the Central Valley. It has been some 25 years since the original RASA ground-water flow model provided valuable insights to the regional ground-water system. Because of the dynamic character of the ground-water flow system, increased geologic and hydrologic knowledge, and the intense competition for available ground-water resources, the USGS re-examined the aquifer beginning in 2004.

Selected Findings

An important step in the USGS's most recent effort to assess the ground-water resources of the Central Valley was to better characterize the hydraulic properties of aquifer sediments. The hydrologic system in the Central Valley is complex, in part because of the heterogeneous nature of the hydrogeologic setting. Fine-grained deposits, including one mappable clay body—the Corcoran Clay—are spread throughout the valley fill. These valley deposits can be conceptualized as one large heterogeneous aquifer.

A database of approximately 8,500 drillers' logs was compiled to organize information on subsurface lithology in the Central Valley. Texture was used as a basis for constructing a three-dimensional spatial correlation model of the percentage of coarse-grained deposits in the Central Valley (fig. 30). This spatial correlation model correlated reasonably well to the geomorphic provinces shown in figure 29. The textural distribution likely is caused by differences in alluvial and fluvial depositional environments in concert with the distribution of the sediment source material. It was further assumed that the textural distribution correlates to hydraulic conductivity, enabling the development of a heterogeneous hydraulic conductivity distribution to represent the framework of the Central Valley ground-water system.

Water development for irrigation has had a pronounced effect on the hydrologic budget of the Central Valley. The development of surface-water and ground-water resources in support of agriculture has fundamentally altered the recharge and discharge components of the Valley's water budget. The USGS Central Valley ground-water availability study developed a water budget for the period 1962–2003, which indicated that irrigation water supply was about evenly split between surface water and ground water with slightly more coming from surface-water delivery (Claudia C. Faunt, U.S. Geological Survey, written commun., 2007). The relative annual contribution from either surface water or ground water, however, can vary substantially in response to climate conditions (wet or dry year). The large amount

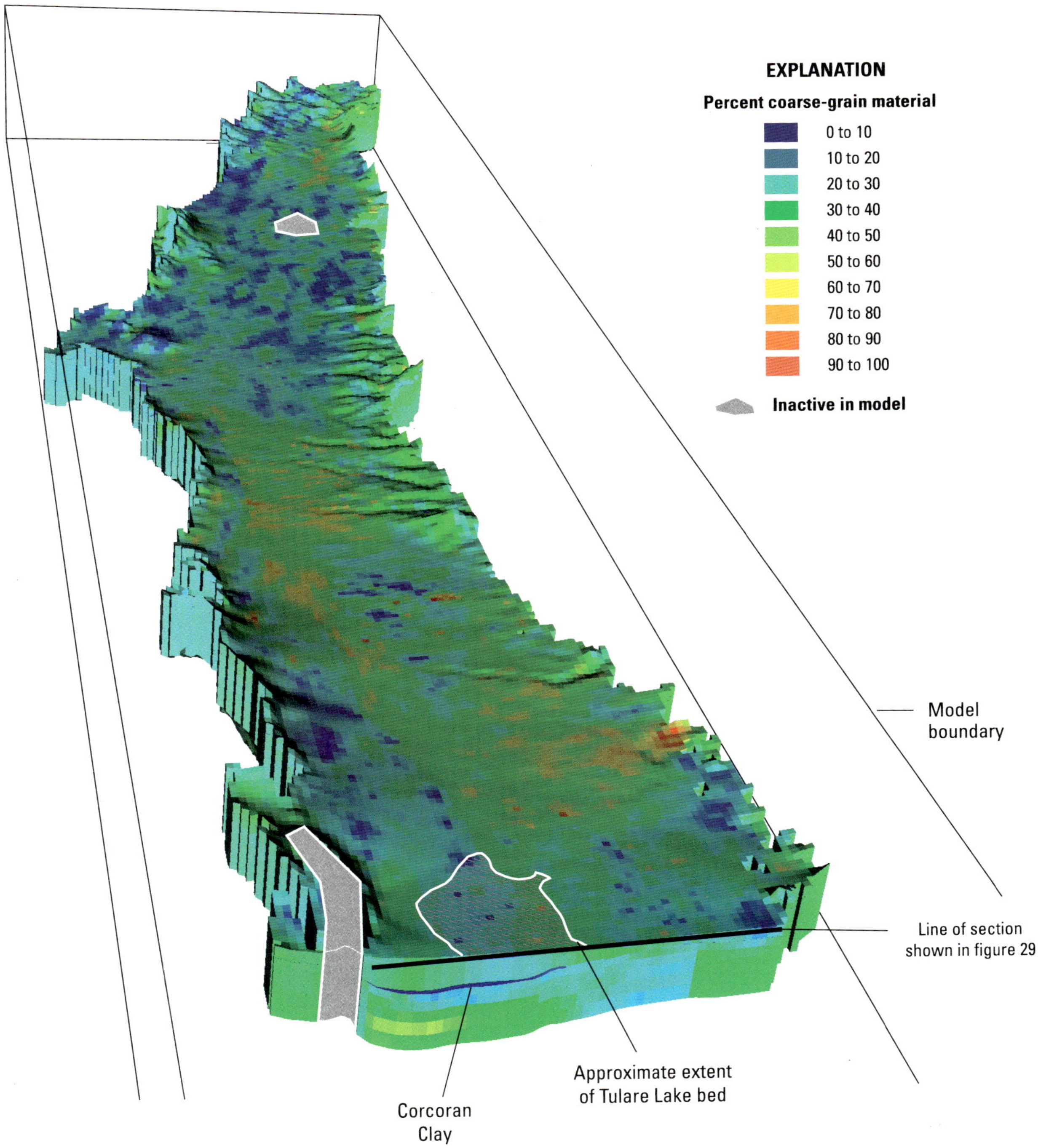

Figure 30. Block diagram of texture model for the Central Valley aquifer (from Claudia C. Faunt, U.S. Geological Survey, written commun., 2007).

of irrigation has altered the amount and distribution of recharge to the system and can change the water-table configuration and volume of surface-water discharge. The predevelopment water budget (fig. 31) indicates that prior to irrigation development, the amount of inflow to the system was equal to outflow (estimated at about 2,800 ft^3/s). For the postdevelopment period (1962–2003), average discharge increased to 15,500 ft^3/s and average recharge increased to 13,500 ft^3/s. Overall, the postdevelopment average discharge and recharge for the aquifer system during this period were about five times greater than the predevelopment values. Postdevelopment average recharge was predominantly from irrigation return flow, and average discharge was overwhelmingly from ground-water pumpage. The average rate of decrease in aquifer storage including subsidence (compaction of fine-grained beds), for the period 1962 through 2003 was estimated to be about 1,900 ft^3/s.

Agricultural development of the Central Valley water resources has resulted in large diversions and redistribution of surface water together with intensive pumpage from wells to meet various crop demands, especially during dry periods. Ground-water pumpage greatly exceeded natural recharge and resulted in large water-level declines. More pronounced in the San Joaquin Valley, but evident elsewhere in the Central Valley, these water-level declines resulted in subsidence. Land subsidence was significant and widespread, causing a variety of practical and economic problems such as damage to canals and drainage systems and loss of irrigation wells. The effects of unchecked water development and the possibility of prolonged drought caused water managers to increase surface-water imports and decrease ground-water pumpage (since the late 1970s). These actions have halted land subsidence in seriously affected areas, but the possibility of subsidence occurring again remains.

The updated ground-water flow model was improved by an increase in its spatial (both laterally and vertically) and temporal resolution, thus enabling enhanced water-budget detail in the upper part of the system. This detail also allows for more accuracy in the estimates of the volume and distribution of subsidence. Perhaps most strikingly, the model now includes the routing of the surface water through the vast series of interconnected rivers and canals, allowing for better simulation of the interaction of ground water and surface water and more detailed water-budget accounting. This is accomplished by using the Farm Process developed for the USGS modular three-dimensional finite-difference ground-water flow model, MODFLOW-2000 (Schmid and others, 2006) linked with the Streamflow-Routing Package (Prudic and others, 2004). As an example, the overall water budget for the final year (2003) of the postdevelopment period comes at the end of 5 years of below-average rainfall. Surface-water agricultural deliveries (13,000 ft^3/s) were more than double the ground-water agricultural deliveries (5,900 ft^3/s) and the resulting combined recharge and surface-water inflow (10,000 ft^3/s) was slightly greater than the combined total agricultural and municipal pumpage (8,000 ft^3/s) for the year 2003 (fig. 31).

An orchard in California's Central Valley. Photograph by Claudia C. Faunt, USGS.

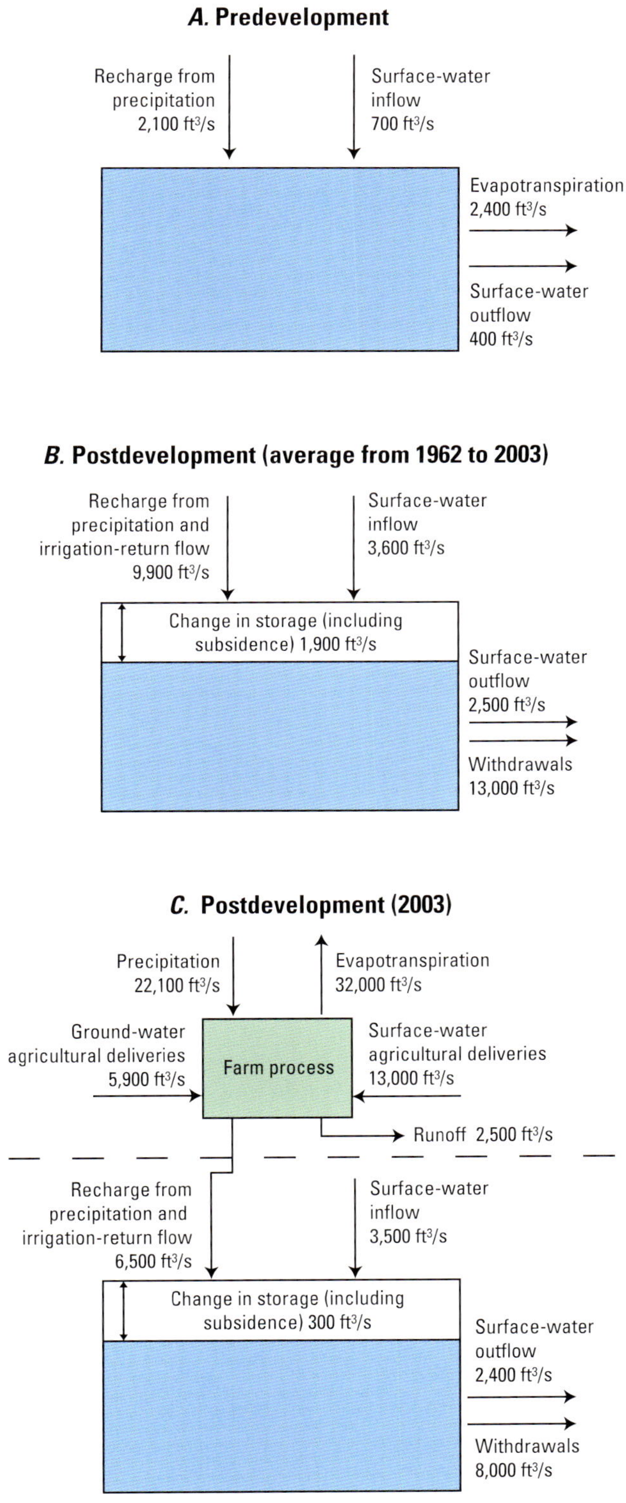

Figure 31. Change in Central Valley aquifer water budget due to development: *(A)* predevelopment (data from Bertoldi and others, 1991), *(B)* postdevelopment (average from 1962–2003) (data from Claudia C. Faunt, U.S. Geological Survey, written commun., 2007), and *(C)* postdevelopment (2003) (data from Claudia C. Faunt, U.S. Geological Survey, written commun., 2007). [ft^3/s, cubic feet per second]

Coastal Plain Aquifer System

Introduction

The study area encompasses the entire Coastal Plain region of North Carolina and South Carolina and extends into parts of Virginia and Georgia (fig. 32). This region is part of the Atlantic and Gulf Coastal Plain physiographic province, which stretches from Long Island, New York, southward and westward into Texas. The portion of the aquifer system in North and South Carolina covers approximately 42,500 mi^2 and is a significant source of water used to meet the needs of a growing population. Most of this area is underlain by at least one, and in many cases several, highly productive aquifers. Water from these aquifers is used for diverse needs, including municipal water supply, as well as industrial and agricultural uses. Many places within this part of the aquifer system have intensively developed the ground-water resource, while in other areas the resource remains largely undeveloped. Some municipalities, where development of the aquifers in previous years created large areas of water-level decline in the aquifer potentiometric surfaces, have switched to surface-water sources, allowing the ground-water levels to partially recover. Saltwater encroachment is occurring in several places within the aquifer system and is directly related to the lowering of the potentiometric surfaces. Most of the North and South Carolina aquifers in the Coastal Plain contain freshwater (total dissolved solids less than 1,000 milligrams per liter), however, brackish water and saltwater exist in the aquifer system, especially near the coast.

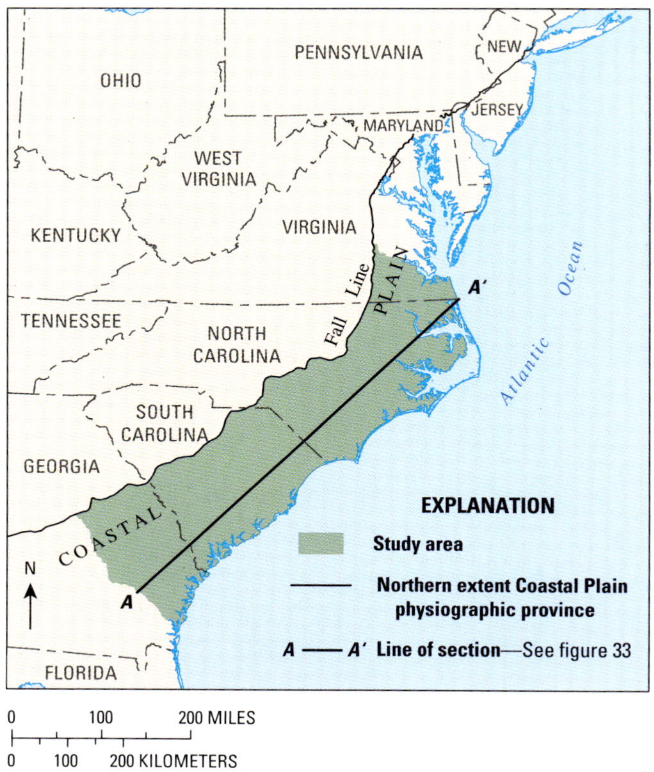

Figure 32. The areal extent of the Coastal Plain aquifer system study area within the Coastal Plain physiographic province and location of cross section (from Bruce G. Campbell, U.S. Geological Survey, written commun., 2007).

The sediments that compose the Coastal Plain range in age from Cretaceous to Holocene (Miller, 1992; Trapp and Horn, 1997). These sediments typically were deposited in shallow marine environments during a series of transgressions and regressions of the Atlantic Ocean. The aquifer system consists of a variety of sediment types from sand, silt, and clay to various types of consolidated carbonate rocks (fig. 33). The boundary to the east of the Coastal Plain is the Atlantic Ocean and the boundary to the west of the Coastal Plain is the Piedmont Province. This boundary separating the Coastal Plain deposits from the Piedmont soils is also called the Fall Line. Coastal Plain sediments thicken from the Fall Line toward the Atlantic Ocean and in many cases extend many miles offshore beneath the sea floor.

The climate of the study area is temperate and characterized by hot, humid summers and moderate winters. Annual precipitation in the area can range from approximately 35 to 65 inches but averages about 50 inches across the two-State area. The water table commonly is near land surface, but a thick (as much as 200 ft) unsaturated zone exists in parts of the upper aquifers. The overall direction of ground-water flow is downdip toward the Atlantic Ocean. Locally, the surface-water features can influence flow directions. Aquifers primarily are recharged by precipitation in the outcrop areas along the Fall Line. Discharge is primarily to nearby streams through the shallow flow system, to other aquifers by leakage, to the ocean, and to wells.

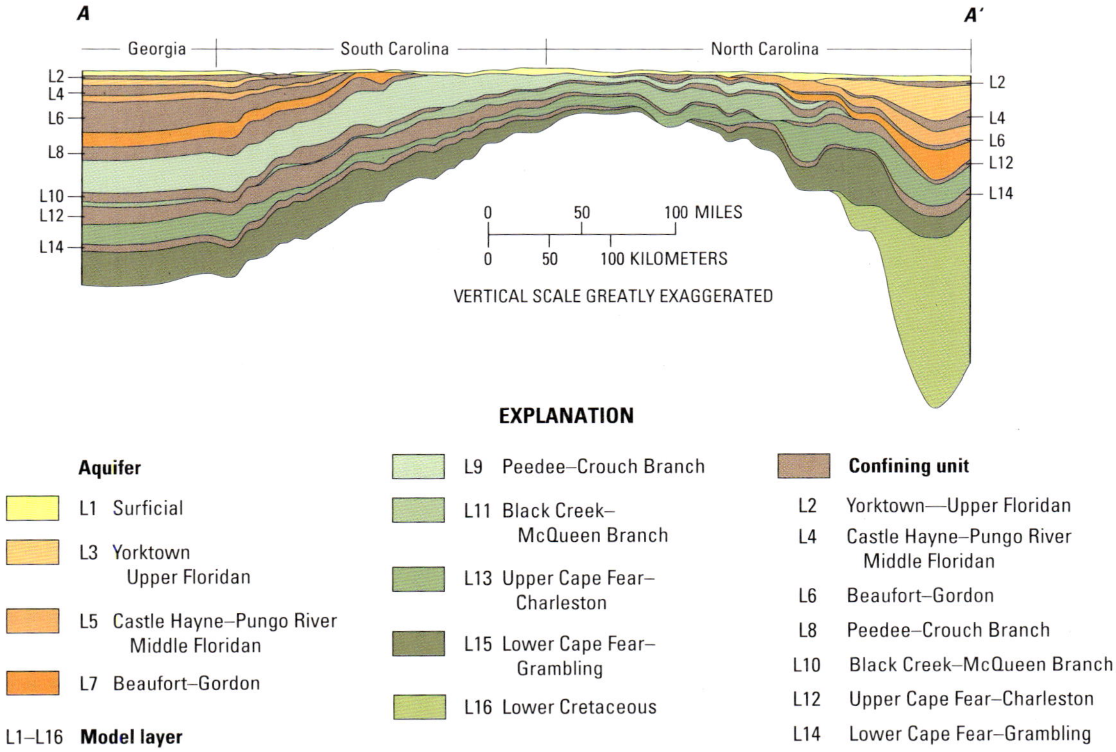

Figure 33. Generalized cross section of hydrogeologic units through the central part of the Coastal Plain aquifer system study area (from Bruce G. Campbell, U.S. Geological Survey, written commun., 2007). Line of section shown in figure 32.

Water Availability Issues

Ground-water withdrawals from the aquifers along the coast in North Carolina and South Carolina have increased substantially over the past couple of decades in response to demands for water for a rapidly growing population. Both States have sought to increase their development of surface-water supplies to meet the needs of the coastal populations but also recognize that additional information is needed about the availability of ground-water resources. For example, the effects of ground-water withdrawals on the quantity of freshwater discharge to streams, estuaries, and wetlands are largely unknown. A complicating factor in some areas is the concern about saltwater intrusion. Adequate ground-water supplies and declining water levels in the study area aquifers of North Carolina and South Carolina became a problem in the late 20th century. Water-level declines have brought attention to the need to better manage withdrawals to maintain ground-water availability for the future. Large, regional areas of water-level decline currently (2007) are present in several areas and are as much as 100–200 ft deep. When these large water-level declines occur near the coast, saltwater intrusion into coastal aquifers is a possibility. Drought compounds the problem of declining water levels. In response to the recent drought of 1998–2002 across the southeastern United States, near-surface ground-water levels in the aquifer system declined to some of the lowest levels on record.

Approach to Assessment

Separate independent numerical models were developed to simulate the ground-water flow in the North Carolina (Giese and others, 1997) and South Carolina (Aucott 1988; 1996) Coastal Plain aquifer system as part of the USGS RASA Program. The North Carolina RASA model simulated conditions from 1900 to 1980, based on a hydrogeologic framework composed of 10 aquifers and associated confining units. The South Carolina RASA model simulated conditions from 1935 to 1982, based on a hydrogeologic framework composed of five aquifers and associated confining units. These inconsistent conceptualizations of the aquifer system and resultant ground-water flow models from adjoining States are now more than 25 years old.

The need to address the study area's ground-water-supply issues and identified differences in the conceptual models of the hydrogeologic framework led to the most recent assessment of ground-water availability that began in 2004. The current regional ground-water flow model took advantage of the large amount of new hydraulic, geologic, water-level, and water-use data available in conducting the hydrologic analysis. USGS scientists constructed a flow model using the updated hydrogeologic framework that was developed in conjunction with State agency partners. This regional aquifer assessment benefited from a variety of investigations conducted by local organizations, State agencies, and private groups. The updated regional model of the aquifer system was calibrated to ground-water levels and stream discharges for both steady-state (predevelopment) and transient (postdevelopment) conditions. An essential part of the modeling effort is collaboration with cooperators and stakeholders and the formation of several project liaison committees. An advantage of this close working relationship with local partners is the development of a regional management tool that can be used to address interstate water issues that involve several of the southeastern States.

Selected Findings

Water development for an expanding population has had a decided effect on the hydrologic budget of the Coastal Plain aquifer system. The simultaneous development of surface-water and ground-water resources to satisfy the needs of the growing population has altered the recharge and discharge components of the ground-water budget of the area. This alteration can be observed by examining the ground-water budget results from the recent

May River in Beaufort County, South Carolina. Photograph by Thomas A. Abrahamsen, USGS.

modeling study. The water-budget analysis indicates that there is an increase in the capture of surface water in the upper aquifers with both an increase in leakage into the aquifer system and a decrease in discharge out of the aquifer system (Bruce G. Campbell, U.S. Geological Survey, written commun., 2007).

The water budget for the aquifer system in the study area was estimated for predevelopment and postdevelopment conditions using the regional ground-water flow model. The results are shown in figure 34. Prior to ground-water development, the amount of inflow to the system (estimated at about 1,920 ft^3/s) was equal to outflow from the system. For the simulated postdevelopment period of 1999 to 2004, average discharge (including pumpage) increased to 2,260 ft^3/s, with recharge from precipitation (1,520 ft^3/s) averaging less than

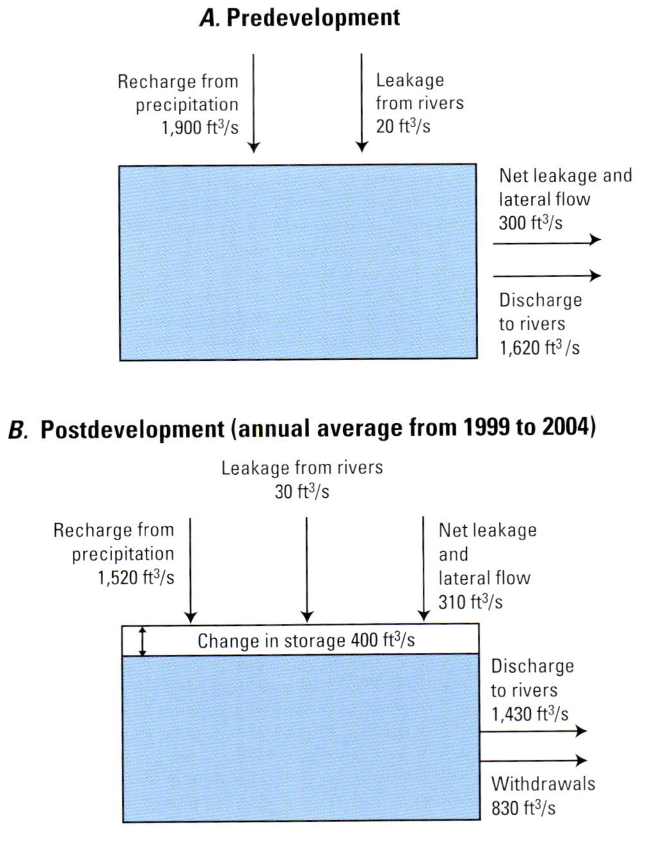

Figure 34. Change in Atlantic Coastal Plain aquifer system water budget due to development: *(A)* predevelopment, and *(B)* postdevelopment (Annual average from 1999 to 2004). (data from Bruce G. Campbell, U.S. Geological Survey, written commun., 2007). [ft^3/s, cubic feet per second]

predevelopment amounts because of drought conditions during the period. This combination has resulted in measured declines in water levels where pumping is most active and where additional surface water has not significantly augmented ground-water withdrawals in meeting the overall water demand. These changes in water levels are the result of an average rate of 400 ft^3/s of water being removed from storage over the 1999 to 2004 postdevelopment period. Over the past couple of decades, the system has recovered in some areas because of the augmentation of ground-water pumpage with surface-water supplies.

It is also important to recognize that the net leakage and lateral flow of the system changed direction between predevelopment and postdevelopment. In predevelopment, a net rate of 300 ft^3/s discharged to the overlying surficial aquifer, the ocean, and the lateral boundaries. For the 1999–2004 period, however, an average net rate of 310 ft^3/s recharged the aquifer system from the overlying surficial aquifer and lateral boundaries. The flows from "net leakage and lateral flow" represent the sum of leakage into the aquifer system from the surficial aquifer being represented as a source of water, leakage out of the system into the surficial aquifer, discharge into the surrounding saltwater bodies, and flow into and out of the modeled area from lateral boundaries. As the system underwent development and water levels declined, the flow from the surficial aquifer into the underlying aquifer system increased, and discharge from the aquifer system into the surficial aquifer and the surrounding saltwater bodies decreased; it is merely a coincidence that this net leakage is about the same amount but in different directions for the predevelopment and the postdevelopment conditions.

Development of the water resources has resulted in localized declines in water levels caused by intensive pumpage from wells primarily to meet public supply demands, which is exacerbated during extended dry periods. When these localized areas of water-level decline occur near the coast, there can be an increased potential for the occurrence of saltwater encroachment. The effects of intense localized ground-water development and climate variability in the form of drought have caused water-resource managers to secure increased surface-water resources and decrease or make plans to decrease ground-water pumpage in selected areas.

Great Lakes Basin

Introduction

The Great Lakes Basin straddles the United States–Canadian border (fig. 35) and covers an area of approximately 296,000 mi^2, including the surface area of the Great Lakes, which is about 30 percent of the total basin area. The Great Lakes watershed includes parts of eight States (Minnesota, Wisconsin, Illinois, Indiana, Michigan, Ohio, Pennsylvania, and New York) in the United States and part of the Canadian province of Ontario. About one-third of the basin land area lies within the United States.

The climate of the Great Lakes Basin is prone to extremes. In winter, cold arctic air moves across the open water of the lakes and absorbs moisture, which is released as snowfall on the leeward side of the lakes as the air mass cools when approaching land. In the summer, the area is dominated by warm, humid air from the Gulf of Mexico (Hodgkins and others, 2007). Mean annual precipitation ranges from less than 27 inches in the west to more than 47 inches in the east. Areal recharge from precipitation is the primary source of water to the ground-water system; with local recharge rates ranging from less than 1 inch to more than 20 inches per year (Neff and others, 2006). Ground water discharges directly to the Great Lakes as seepage or indirectly as base flow in streams and rivers.

The amount of water flowing through the subsurface varies, depending on the characteristics of the water-bearing rocks and sediments. Unconsolidated materials that were deposited at or near land surface as a result of large-scale glacial advances and retreats make up the most productive aquifers. Most glacial deposits are composed of mixtures of sand and gravel, and silt and clay. Glacial deposits can be as much as 1,200 ft thick in parts of Michigan and are several hundred feet thick in buried bedrock valleys in Illinois, Wisconsin, and New York (Grannemann and others, 2000). Deposits can be thin or nonexistent in areas where bedrock is exposed at land surface due to removal of material by glaciers.

Bedrock aquifers are generally widespread throughout the region and are more continuous than the aquifers in the glacial deposits (fig. 36). Some bedrock aquifers extend beyond the watershed boundaries. The extent and

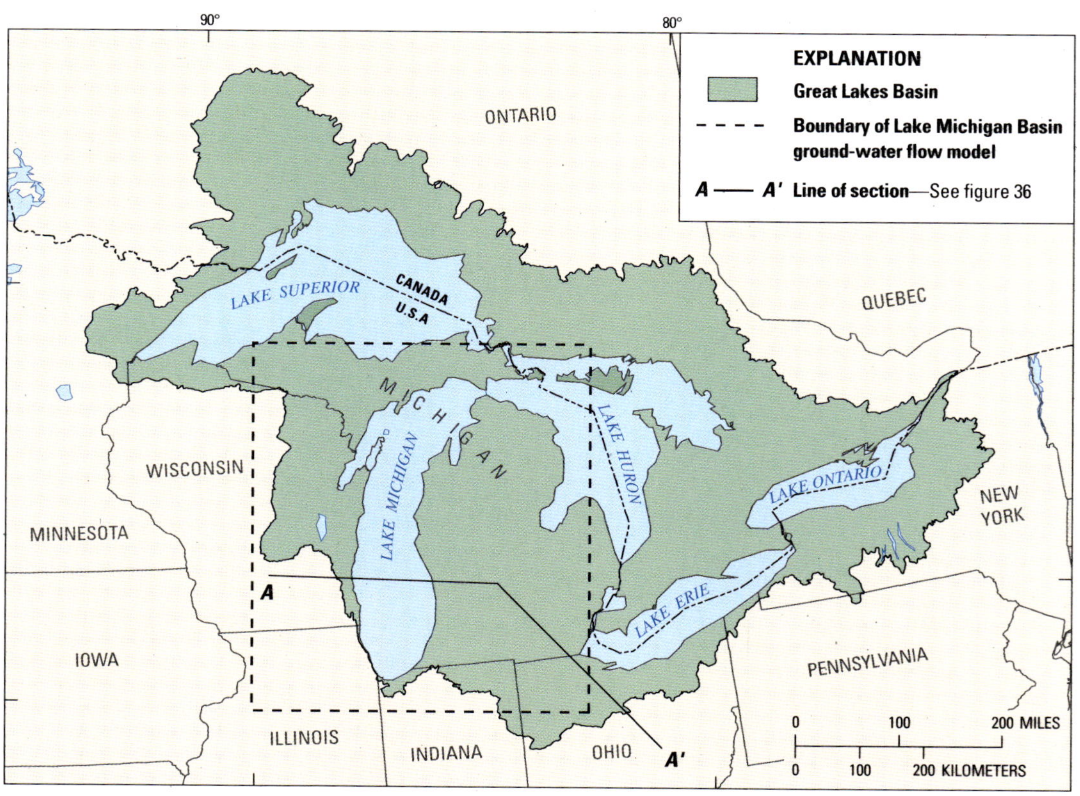

Base from ESRI, 2001; U.S. Army Corps of Engineers, 1998; and Environment Canada, 1995

Figure 35. The Great Lakes Basin in the United States and Canada.

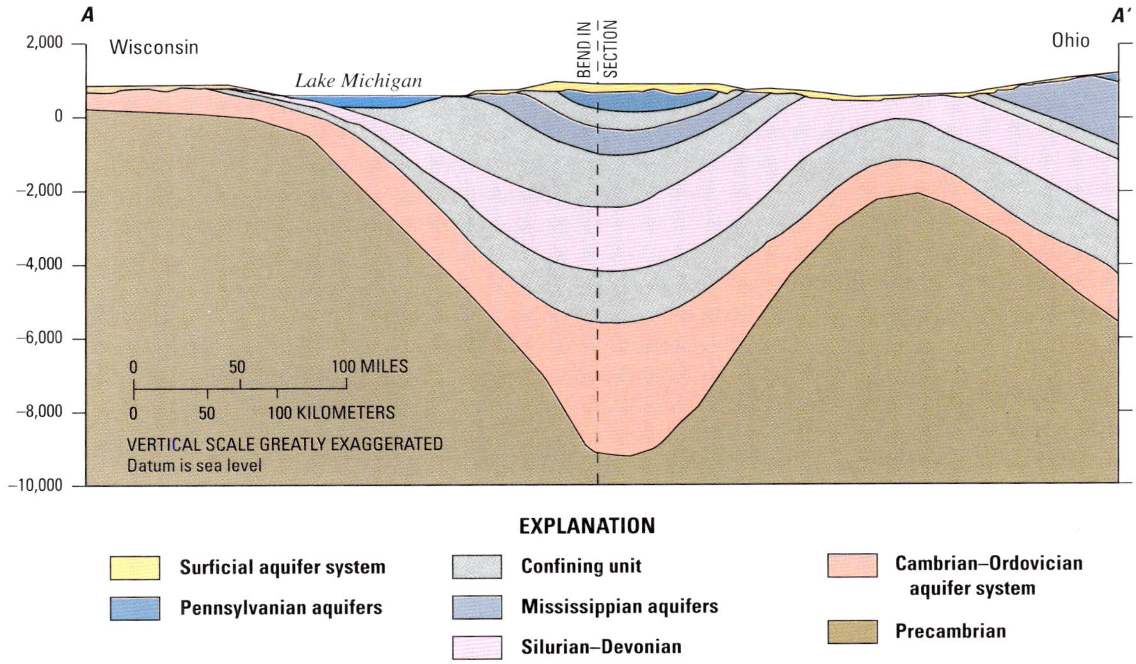

Figure 36. Generalized hydrostratigraphic section through the central part of the Great Lakes Basin, United States (from Sheets and Simonson, 2006). Line of section shown in figure 35.

boundaries of the ground-water basin may be different than the surface expression of watersheds (ground-water divides and watershed boundaries may not coincide). Additionally, as the depth to these aquifers increases so does the likelihood that water quality will degrade and the less likely the aquifers will be used for water supply.

Water Availability Issues

Most large public water supplies in the basin are taken directly from the Great Lakes. Ground water has always been an important source of drinking water in rural areas; however, it is becoming an increasingly important source of drinking water (supplying about 8.2 million people within the watershed) as the population spreads beyond the large municipalities into the suburbs (Grannemann and others, 2000). In the past, agriculture alone dominated rural water use, but competing demands have resulted in the conversion of some areas of agricultural water use to drinking water supplies. With continued population growth come additional demands for ground water to satisfy other uses such as manufacturing, power generation, and transportation. Ground-water availability is directly linked to the suitability of the ground-water quality for its intended use. From a ground-water perspective, major resource issues related to water availability are focused around finding adequate supplies, understanding the relation between surface water and ground water, the effect of changing water quality on water use, and ecosystem health in relation to quantity and quality of the water resource. The issue of diverting water from the Great Lakes remains a keen interest.

Approach to Assessment

In 2005, the USGS began a pilot effort in the Great Lakes Basin to develop a potential nationwide program that would describe the status and trends in the availability and use of the Nation's freshwater resources (Barlow and others, 2002; Grannemann and Reeves, 2005). Prior to this most recent endeavor, the ground-water resources of the basin were comprehensively assessed in the early 1970s (Allen and Waller, 1975). The USGS studies of regional aquifer systems by the RASA Program during 1978–95 contributed greatly to the geologic and hydrologic knowledge of the ground-water resources of the area (Young, 1992; Westjohn and Weaver, 1998; Bugliosi, 1999; Randall, 2001). The Great Lakes Basin pilot study is building on this information together with other available information on surface-water flows and storage of water.

Selected Findings

The Great Lakes Basin pilot study is at its halfway point (2007); therefore, a basinwide assessment of water availability is incomplete. However, several interim findings are available that improve understanding of the most important factors affecting the Great Lakes Basin hydrologic system now and for the future. Initially, to improve understanding of water resources in the Great Lakes, it is important to identify the role ground water plays in the hydrologic budget (Grannemann and others, 2000). In estimates of sources of water to the Great Lakes, direct ground-water discharge generally is assumed to be small. This perspective, however, does not account for the amount of ground water that flows indirectly into the Great Lakes as a component of streamflow. A recent study by Neff and others (2005) analyzed streamflow-monitoring stations across the basin and determined that the ground-water component of streamflow averages 66 percent of total streamflow for the basin and is a significant source of water to the Great Lakes.

The location of ground-water divides determines whether ground water flows toward or away from one of the Great Lakes (Sheets and Simonson, 2006). In the shallow ground-water system, ground-water divides typically coincide approximately with watershed divides. In deeper aquifers, however, this is not necessarily the case. Additionally, the location of ground-water divides can change if the system is stressed, whether naturally (drought) or through intense development of the ground-water system.

Consistent and accurate estimates are needed to understand how recharge might affect ground-water availability and use. Some recent progress has been made in estimating recharge on a basinwide scale for the Great Lakes Basin. Neff and others (2006) provide the first integrated study of long-term average ground-water recharge to the shallow aquifers in the United States and Canada within the Great Lakes Basin. Results are limited to long-term averages, broad spatial scales, and fluxes to the shallow (less than 100 ft) system. Additional studies of deep aquifer recharge and the temporal variability of recharge are needed to gain a better understanding of ground-water recharge in the Great Lakes Basin.

In the United States part of the Great Lakes Basin, four regional aquifer-system studies—Northern Midwest, Midwestern Basins and Arches, Northeast Glacial Aquifers, and the Michigan Basin—were conducted by the USGS as part of the RASA Program. The Great Lakes Basin ground-water system (in the United States) is composed of three major aquifer systems: Cambrian–Ordovician aquifer system, the Silurian–Devonian aquifers, and the surficial aquifer system; and three minor aquifer systems: Pennsylvanian sandstone aquifer, the Pennsylvanian sandstone and carbonate-rock aquifer, and the Mississippian sandstone aquifer (U.S. Geological Survey, 2003). Recent examination of regional studies indicates that the regional ground-water divides for the Cambrian–Ordovician aquifer system and the Silurian–Devonian aquifers have changed over time and differ from surface-water divides in some areas (Sheets and Simonson, 2006). These differences are a result of either pumping or structural geologic features and make the assessment of individual water-budget components more difficult and highly dependent on which areas are being evaluated. Currently (2007), the USGS is developing a ground-water flow model of the Lake Michigan ground-water basin (fig. 35). Preliminary results from the steady-state simulation for the water budget indicate that recharge from precipitation accounts for about 90 percent of the inflow to the Lake Michigan Basin (Howard W. Reeves, U.S. Geological Survey, written commun., 2007) (fig. 37). The water budget depicted in figure 37 is for the predevelopment Lake Michigan Basin watershed; the net-lateral inflow component represents the flow in the ground-water system that crosses the topographic watershed boundary.

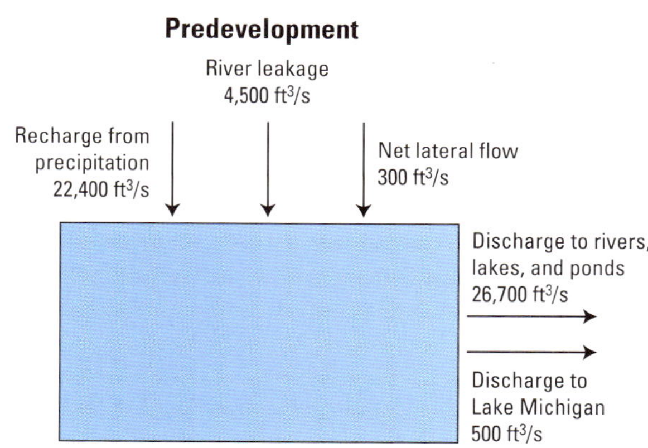

Figure 37. Predevelopment Lake Michigan Basin ground-water budget (data from Howard W. Reeves, U.S. Geological Survey, written commun., 2007). [ft^3/s, cubic feet per second]

High Plains Aquifer

Introduction

The High Plains aquifer is one of the largest and most productive aquifers in the Nation. It underlies an area of about 111 million acres (174,000 mi^2) in parts of eight western States (fig. 38). The area is a remnant of a vast plain formed by sediments that were deposited by streams flowing eastward from the ancestral Rocky Mountains. Mean annual precipitation ranges from 16 inches in the western part of the High Plains to about 28 inches in the east. Evaporation rates measured from free-water surfaces in the High Plains range from 60 inches in the north to 105 inches in the south (Gutentag and others, 1984). These rates are among the highest in the Nation and are attributed to high summer temperatures and persistent winds. Because evaporation rates are high relative to precipitation, there is little water available to recharge the aquifer, and in some areas, the time between recharge events may be years, decades (Luckey and Becker, 1999), or even longer (McMahon and others, 2006).

Depending on location, the High Plains aquifer consists of one or more hydraulically connected geologic units. In most of the area, the Ogallala Formation of Miocene age and overlying hydraulically connected Quaternary deposits, if present, are the principal geologic units of the aquifer (fig. 39). The geologic units that form the High Plains aquifer generally were deposited by streams or wind and contain varying amounts of clay, silt, sand, and gravel. Bedrock units that underlie the High Plains aquifer are for the most part consolidated and are not as transmissive as the sediments that compose the High Plains aquifer. Gutentag and others (1984) provide an overview of the geologic history and setting of the High Plains.

The saturated thickness of the High Plains aquifer ranges from about zero to more than 1,000 ft and averages about 200 ft. Depth to water ranges from near land surface to 500 ft and has an average depth of about 100 ft. The ground-water flow direction generally is from west to east and locally toward streams. Pumping from numerous irrigation wells across the High Plains is the primary mechanism for ground-water discharge. Precipitation is the principal natural source of recharge to the aquifer. Regional variability in water-level changes in the High Plains aquifer results from regional differences in climate, soils, land use, and ground-water withdrawals for irrigation.

Water Availability Issues

The High Plains aquifer is the source of irrigation water that has transformed a part of the Great Plains into one of the major agricultural regions of the world. About 97 percent of the water pumped from the aquifer is used for irrigation, which accounts for about 30 percent of the ground water withdrawn for irrigation in the United States (Maupin and Barber, 2005). Of the total crop production in the United States, the High Plains aquifer area accounts for about 19 percent of the wheat, 19 percent of the cotton, 15 percent of the corn, and 3 percent of the sorghum (U.S. Department of Agriculture, 1999). Irrigation first began in the southern High Plains in the late 1880s, but it was not until after the drought in the 1930s that it became clear that increased development of water for irrigation was necessary. The postwar economic boom of the 1950s, the development of irrigation technologies, and federally funded irrigation projects in the 1960s allowed farmers to work the land and become independent of the natural dry and wet weather cycles that had previously frustrated their efforts to cultivate the plains (Dennehy and others, 2002). By 1992, irrigated crop land (13 million acres) represented about 12 percent of the High Plains land use (Qi and others, 2002). In addition to crops, the region accounts for nearly 18 percent of the total cattle production in the United States and is rapidly becoming a center for swine production. The aquifer also provides drinking water to most of the people who live within its boundaries.

Successful development of the ground-water resources does not come without a cost. Substantial pumping of the High Plains aquifer for irrigation has resulted in water-level declines in some parts of the aquifer of more than 150 ft (McGuire, 2007). In 1984, concern about these declines led Congress to mandate a water-level monitoring program for the aquifer.

Approach to Assessment

To date, there have been three regionally extensive ground-water assessment studies of the High Plains aquifer (Johnson, 1901, 1902; Lohman, 1953; Weeks and others, 1988). The latest and most comprehensive investigation to quantify the ground-water resources was conducted by Weeks and others (1988) as part of the RASA Program to evaluate

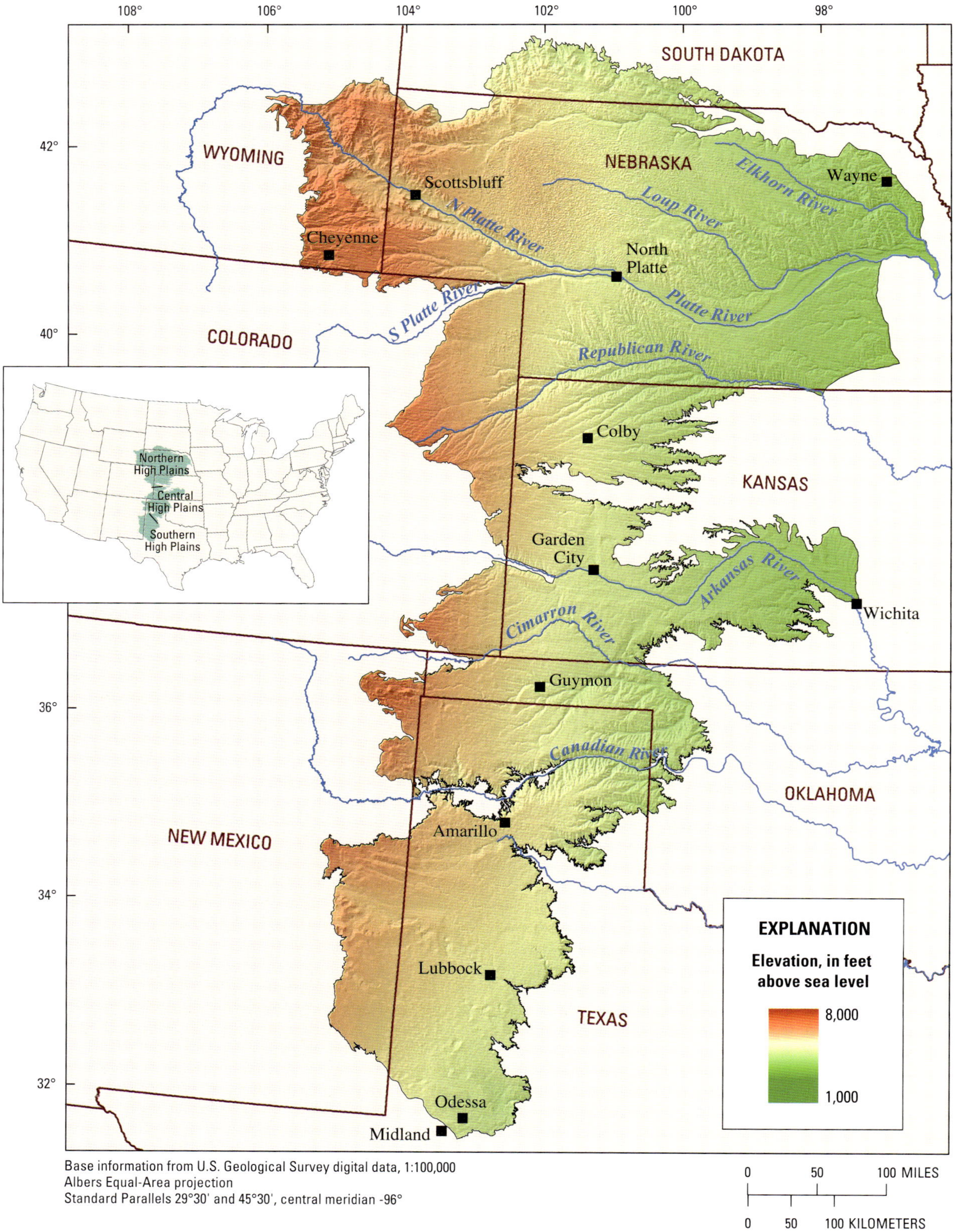

Figure 38. The High Plains aquifer in the west-central United States.

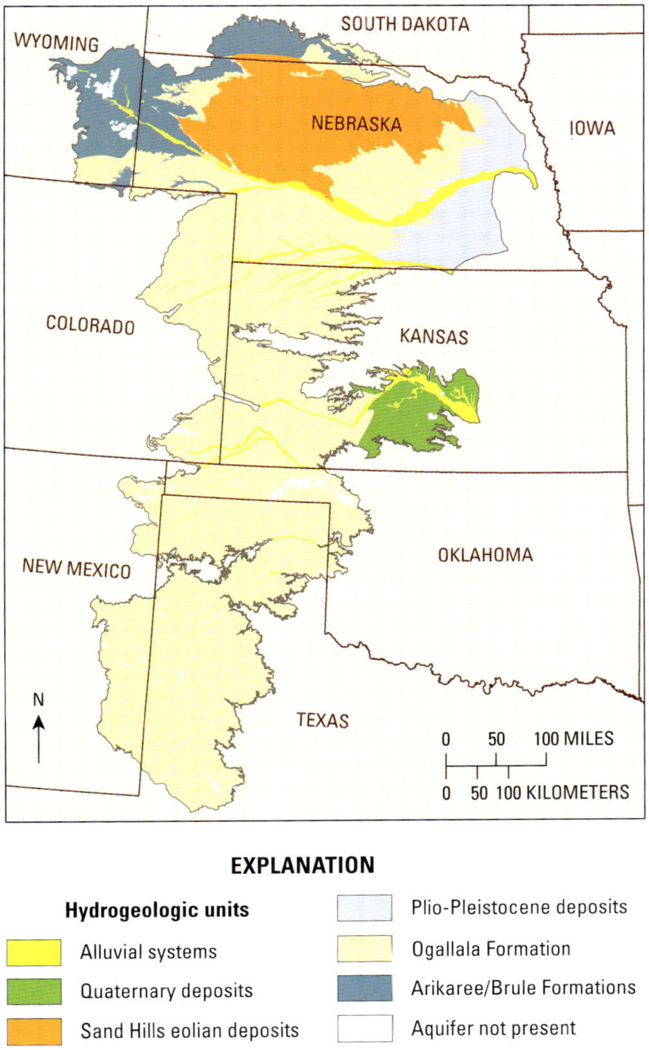

Figure 39. Hydrogeologic units of the High Plains aquifer (modified from Gutentag and others, 1984).

the historical and future effects of ground-water development on the High Plains aquifer. The High Plains RASA assembled extensive amounts of information used to define the aquifer's hydrogeologic framework, various water-quality characteristics, and hydrologic budget and stresses. These data were used to perform a quantitative analysis using the first computer model of the entire aquifer (Luckey and others 1986; Luckey and others, 1988). The computer model enabled investigators to examine the system response to a variety of ground-water development strategies. Intensive ground-water withdrawals continue to support the agricultural lifestyle that has existed in the area since the early 1900s. There have been no system-wide updates to the computer models that were constructed more than 20 years ago.

To better predict the future availability of ground water, it is necessary to improve our understanding of a variety of topics, including ground-water quality, because ground-water quality has a direct effect on how water can be used. The first regionally comprehensive ground-water-quality assessment of the High Plains aquifer began in 1998 as part of the USGS NAWQA Program (Dennehy, 2000; McMahon and others, 2007). The purpose of this study was to determine the current ground-water quality conditions of the High Plains aquifer and to increase the scientific understanding of the factors that affect ground-water quality. The quantity and quality of ground water in the High Plains aquifer need to be considered together when assessing ground-water availability.

Selected Findings

Ground-water withdrawals, especially during the past 50 years, have resulted in significant water-level declines that have adversely affected the long-term sustainability of the High Plains aquifer. Water-level changes in the High Plains aquifer from predevelopment to 2005 are shown in figure 40. About 24 percent of the area or about 27 million acres had more than 10 ft of water-level decline from predevelopment to 2002 (McGuire, 2007). Large areas with more than 50 ft of decline predominate in the southern High Plains and extend north into southwest Kansas. Additional information on historical water levels is available at the High Plains Water-Level Monitoring Study web site (*http://ne.water.usgs.gov/ogw/hpwlms/*). During the past half century or so, water in storage in the High Plains aquifer declined about 200 million acre-feet (equivalent to 65 trillion gallons) with 62 percent of the total ground-water volume loss occurring in Texas. Although the rate at which water levels are declining has slowed, the downward trend continues in many areas across the High Plains (McGuire, 2007).

The water used for irrigation is not coincident with the largest reservoir of water in storage. Ground-water withdrawals initially began in the southern High Plains and moved northward. The largest drawdowns in the High Plains aquifer are contained within the southern High Plains. Therefore, it is misleading to look at a water budget for the entire High Plains when the southern half of the aquifer has undergone greater withdrawals from a smaller reservoir of available ground water. Reduction in storage primarily due to ground-water development for irrigation is responsible for about a 7-percent reduction of the original total water volume when considering the entire High Plains aquifer. If we compare water budgets developed for just the south-central High Plains aquifer for predevelopment and for 1997, however, an 18-percent decline of water in storage is computed for that area

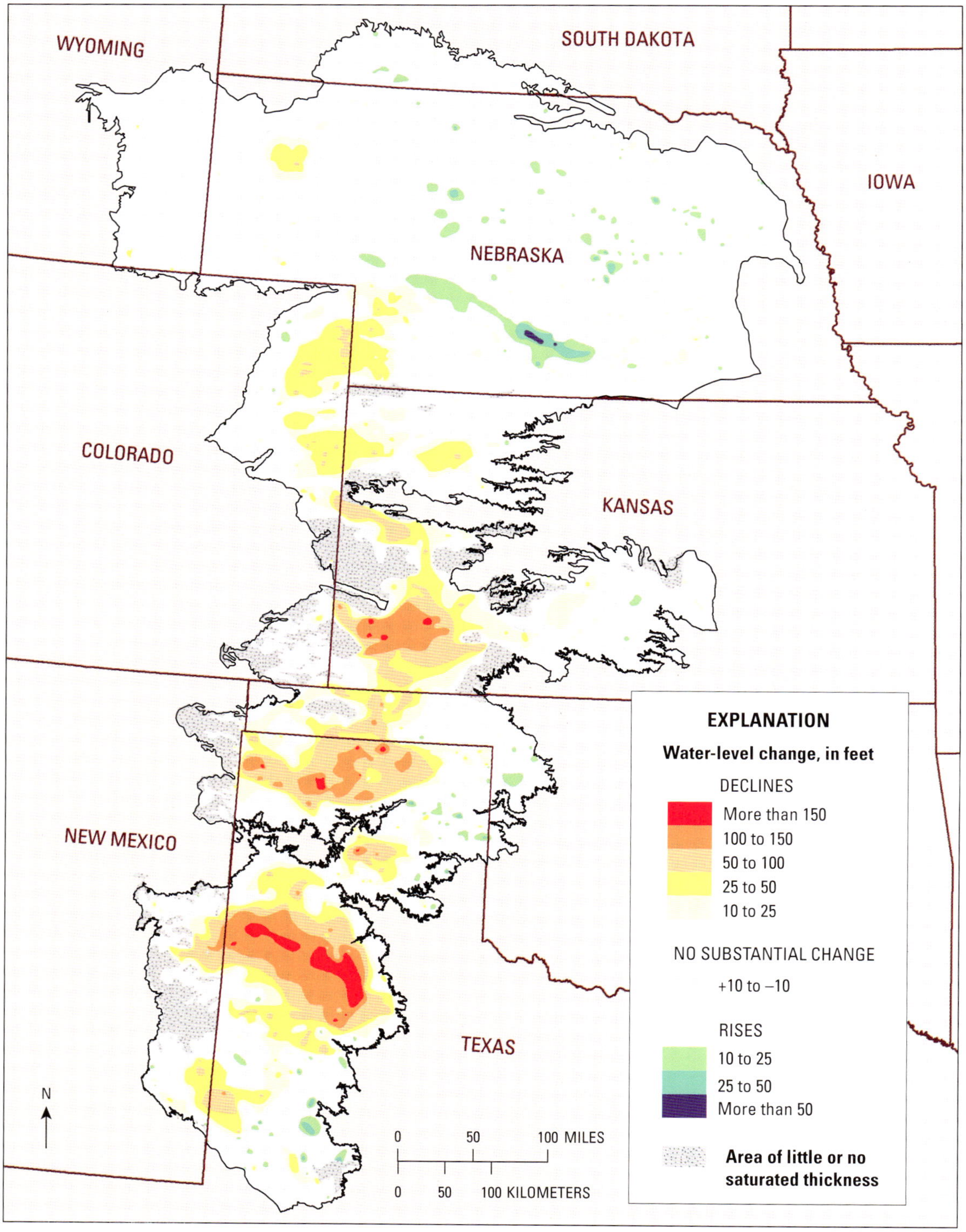

Figure 40. Water-level changes in the High Plains aquifer, predevelopment to 2005 (modified from McGuire, 2007).

(Luckey and Becker, 1999). As shown in figure 41, more than 80 percent of the ground-water withdrawals at the end of 1997 in the south-central High Plains aquifer comes from ground-water storage.

Water in the High Plains aquifer generally is of high quality and is suitable for most agricultural, domestic, and industrial uses without treatment. The intensive agricultural land use and large-scale changes in the hydrologic regime, however, potentially affect the availability of the ground-water resource through changes in water quality. Transport of human and agricultural chemicals (nitrogen and pesticides) in ground-water recharge from land surface to the water table is perhaps the most obvious way in which ground-water quality has been degraded, but other factors should be considered. For example, infiltration of water from multiple irrigation applications can increase the concentration of dissolved solids through evaporation. Likewise, conversion of natural rangeland to agricultural lands can adversely affect shallow ground-water quality by changing local recharge and flushing salts that have accumulated at the surface over long periods of time to underlying aquifers (Scanlon and others, 2005; McMahon and others, 2006). Intensive pumping and resulting declines in water levels can enhance the upward movement of mineralized water from below by increasing upward hydraulic gradients into the aquifer. The combination of degraded water from above and mineralized ground water from below can limit the unaffected water to a layer in the center of the aquifer that has not been affected by human activities. A confounding situation can exist, however, if the density of high-capacity pumping wells is sufficient enough to mix the water in the aquifer and thus begin producing water the quality of which is a blend of the conditions across the saturated thickness. This situation is illustrated using ground-water ages in figure 42. Ground-water ages typically increase exponentially with depth, which can result in a relatively thin layer of modern recharge overlying a much thicker zone of old recharge; whereas, in the second graph of figure 42 ground-water ages do not change with depth because of vertical mixing caused by pumping.

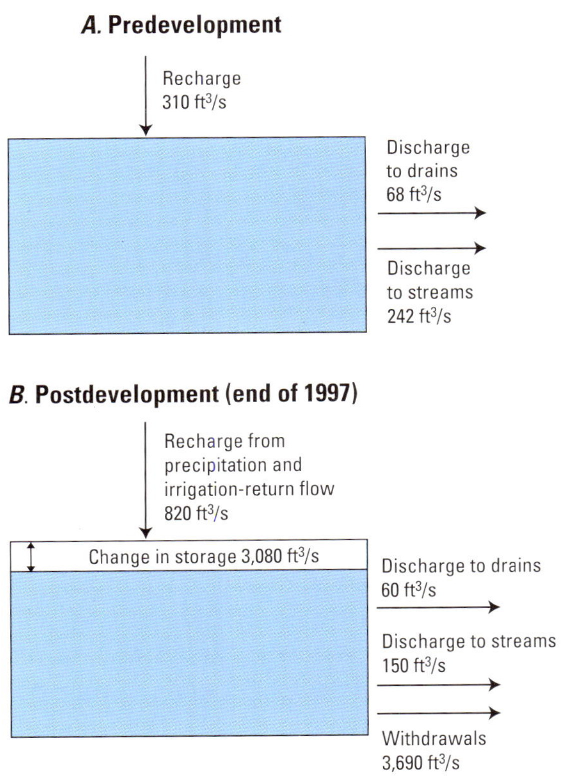

Figure 41. Ground-water budget in the south-central High Plains aquifer study area: *(A)* predevelopment and *(B)* at the end of 1997 (data from Luckey and Becker, 1999). [ft³/s, cubic feet per second]

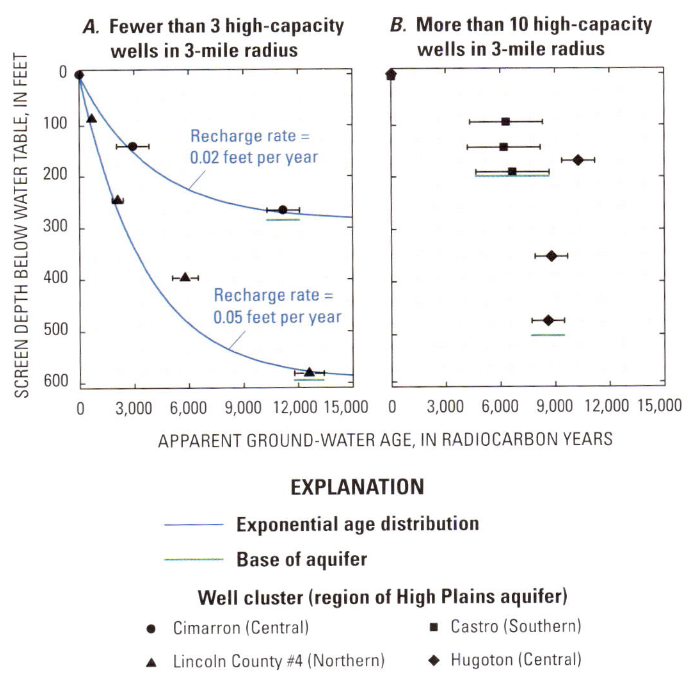

Figure 42. Apparent ground-water age and depth below the water table in the High Plains aquifer: *(A)* flow system relatively undisturbed by pumping and *(B)* flow system highly perturbed by pumping (McMahon and others, 2007).

Future Directions

Ground-water systems change in response to development and changes in climate, and need to be monitored and evaluated on a regular basis to quantify the amount of water available for use and the ramifications of using the resource. Each regional ground-water system is unique in terms of climate, hydrogeologic framework, and boundary conditions (both type and location), and each system responds differently to stresses from human development and climate. For example, the five case studies presented here illustrate how water budgets and individual budget components differ geographically and demonstrate how human intervention has influenced water movement in several major aquifer systems.

> *"Continue to conduct research and provide information—at a scale that is useful to states and local entities—about such matters as the safe, or sustainable, yield of aquifers (and methods for determining that yield); water-use data; and delineating boundaries and water budgets of three-dimensional watersheds, including scientifically based and cost-effective methods of quantifying interactions between ground water and surface water."*
>
> Recommended actions to USGS and State Geological Surveys from "Ground Water Report to the Nation," Ground Water Protection Council (2007)

An approach for broad-scale assessment of the Nation's ground-water resources is proposed that is adaptable over time and provides quantitative regional analyses of major areas of ground-water use. The proposed program builds on the results of a long history of partnerships among the USGS, other Federal agencies, States, Tribes, and local governments to collect ground-water data and undertake investigative studies of ground-water systems. The proposed program would provide current estimates and historic trends in ground-water use, storage, recharge, and discharge (water-budget analysis); computer models of regional ground-water systems; regionwide estimates of hydraulic properties for major aquifers; evaluation of existing networks for monitoring ground-water availability; and testing and evaluating new approaches for analysis of regional aquifers.

The proposed program is designed to allow both "scaling up" to a national synthesis and "scaling down" to provide information relevant to issues of more local concern. With respect to scaling up, the regional studies undertaken by the program together with additional information from the USGS and other agencies could be synthesized periodically to provide to the extent possible a comprehensive up-to-date picture of the Nation's ground-water resources. Such information on aquifer conditions and long-term factors affecting ground-water resources would help address broad-based societal issues related to present day availability and long-term sustainability of the ground-water resources that are essential to our Nation's water supply, agricultural and industrial production, and ecosystem health.

With respect to scaling down, it is recognized that ground-water management decisions in the United States are made by States, municipalities, and special districts formed for ground-water management. Thus, information and models provided at the regional scale would be designed to provide a regional framework for more detailed studies and models by those who make management decisions at the local level. In addition, the regional studies would be partnered, where possible, with interested agencies and organizations to enhance their relevance to local concerns.

Acknowledgments

The assistance of Eve L. Kuniansky, Devin L. Galloway, Dennis W. Risser, and Patrick Tucci of the USGS in assembling table 2 is gratefully acknowledged. The ground-water-level decline data from the USGS National Water Information System were compiled with the assistance of George A. Karavitis. The reviews by Mark T. Anderson, James R. Bartolino, James W. LaBaugh, and Keith R. Prince improved the technical content of the entire report. Review of individual case studies and provision of additional unpublished information by Bruce G. Campbell, Claudia C. Faunt, Douglas P. McAda, Peter B. McMahon, and Howard W. Reeves of the USGS improved the content and accuracy of the summaries. The editorial assistance of Kay E. Hedrick and Sandra C. Cooper is also gratefully acknowledged. Sharon L. Qi provided the Geographical Information System support. And lastly, the final report preparation and layout by Caryl J. Wipperfurth is especially appreciated.

Selected References*

Ackerman, D.J., 1996, Hydrology of the Mississippi River Valley alluvial aquifer, South-Central United States: U.S. Geological Survey Professional Paper 1416-D, 56 p., 8 pls.

Allen, W.B., and Waller, R.M., 1975, Great Lakes Basin framework study: Appendix 3—Geology and ground water: Great Lakes Basin Commission, Ann Arbor; Appendix 3 of Great Lakes Basin Framework Study, 152 p.

Alley, W.M., 2006, Tracking U.S. groundwater—Reserves for the future?: Environment, v. 48, no. 3, p. 10–25.

Alley, W.M., 2007, Another water-budget myth—The significance of recoverable ground water in storage: Ground Water, v. 45, no. 3, p. 251.

Alley, W.M., Healy, R.W., LaBaugh, J.W., and Reilly, T.E., 2002, Flow and storage in groundwater systems: Science, v. 296, no. 5575, p. 1985–1990.

Alley, W.M., and Leake, S.A., 2004, The journey from safe yield to sustainability: Ground Water, v. 42, no. 1, p. 12–16.

Alley, W.M., Reilly, T.E., and Franke, O.L., 1999, Sustainability of ground-water resources: U.S. Geological Survey Circular 1186, 79 p.

Anderholm, S.K., 2000, Mountain-front recharge along the eastern side of the Middle Rio Grande Basin, central New Mexico: U.S. Geological Survey Water-Resources Investigations Report 00–4010, 36 p.

Anderson, M.T., and Woosley, L.H., Jr., 2005, Water availability for the Western United States—Key scientific challenges: U.S. Geological Survey Circular 1261, 85 p.

Anderson, T.W., 1995, Summary of the Southwest Alluvial Basins, Regional Aquifer-System Analysis, south-central Arizona and parts of adjacent States: U.S. Geological Survey Professional Paper 1406-A, 33 p.

Arthur, J.K., and Taylor, R.E., 1998, Ground-water flow analysis of the Mississippi embayment aquifer system, South-Central United States: U.S. Geological Survey Professional Paper 1416-I, 48 p., 9 pls.

Ator, S.W., Denver, J.M., Krantz, D.E., Newell, W.L., and Martucci, S.K., 2005, A surficial hydrogeologic framework for the Mid-Atlantic Coastal Plain: U.S. Geological Survey Professional Paper 1680, 44 p.

Aucott, W.R., 1988, The predevelopment ground-water flow system and hydrologic characteristics of the Coastal Plain aquifers of South Carolina: U.S. Geological Survey Water-Resources Investigations Report 86–4347, 66 p.

* Most of the USGS reports are available online at *http://pubs.usgs.gov*

Aucott, W.R., 1996, Hydrology of the southeastern Coastal Plain aquifer system in South Carolina and parts of Georgia and North Carolina: U.S. Geological Survey Professional Paper 1410-E, 83 p.

Barlow, P.M., 2003, Ground water in freshwater-saltwater environments of the Atlantic Coast: U.S. Geological Survey Circular 1262, 113 p.

Barlow, P.M., and others, 2002, Concepts for national assessment of water availability and use: U.S. Geological Survey Circular 1223, 34 p.

Bartolino, J.R., and Cole, J.C., 2002, Ground-water resources of the Middle Rio Grande Basin, New Mexico: U.S. Geological Survey Circular 1222, 132 p.

Bartolino, J.R., Cole, J.C., and Hester, D.J., 2002, Ground-water resources of the Middle Rio Grande Basin: U.S. Geological Survey Fact Sheet 088–02, 6 p.

Belcher, W.R., ed., 2004, Death Valley regional ground-water flow system, Nevada and California—Hydrogeologic framework and transient ground-water flow model: U.S. Geological Survey Scientific Investigations Report 2004–5205, 408 p., 2 pls.

Bentley, H.W., Phillips, F.M., Davis, S.N., Habermehl, M.A., Airey, P.L., Calf, G.E., Elmore, D., Grove, H.E., and Torgersen, T., 1986, Chlorine 36 dating of very old groundwater—1. The Great Artesian Basin, Australia: Water Resources Research, v. 22, p. 1991–2001.

Berkstresser, C.F., 1973, Base of fresh ground water, approximately 3,000 micromhos, in the Sacramento Valley and the Sacramento–San Joaquin Delta, California: U.S. Geological Survey Water-Resources Investigations 73–40, 1 sheet.

Bertoldi, G.L., Johnston, R.H., and Evenson, K.D., 1991, Ground water in the Central Valley, California—A summary report: U.S. Geological Survey Professional Paper 1401-A, 44 p.

Bredehoeft, John, 2003, From models to performance assessment—The conceptualization problem: Ground Water, v. 41, no. 5, p. 571–577.

Bredehoeft, John, 2007, It is the discharge: Ground Water, v. 45, no. 5, p. 523.

Bugliosi, E.F., 1999, The Midwestern Basins and Arches regional aquifer system in parts of Indiana, Ohio, Michigan, and Illinois; summary: U.S. Geological Survey Professional Paper 1423-A, 46 p.

Bureau of Reclamation, 1994, The Central Valley project overview, Bureau of Reclamation History Program, Research on Historic Reclamation Projects; accessed November 20, 2007, at *http://www.usbr.gov/dataweb/html/cvpintro.html*

Burkart, M.R., 1984, Availability and quality of water from the Dakota aquifer, northwest Iowa: U.S. Geological Survey Water-Supply Paper 2215, 65 p., 8 pls.

Burow, K.R., Shelton, J.L., Hevesi, J.A., and Weissmann, G.S., 2004, Hydrogeologic characterization of the Modesto area, San Joaquin Valley, California: U.S. Geological Survey Scientific Investigations Report 2004–5232, 54 p.

Chichester, D.C., 1996, Hydrogeology of, and simulation of ground-water flow in, a mantled carbonate-rock system, Cumberland Valley, Pennsylvania: U.S. Geological Survey Water-Resources Investigations Report 94–4090, 39 p.

Clark, D.W., and Kendy, Eloise, 1992, Plan of study for the Regional Aquifer-System Analysis of the northern Rocky Mountains intermontane basins, Montana and Idaho: U.S. Geological Survey Water-Resources Investigations Report 92–4116, 16 p.

Conlon T.D., Wozniak, K.C., Woodcock, D., Herrera, N.B., Fisher, B.J., Morgan, D.S., Lee, K.K., and Hinkle, S.R., 2005, Ground-water hydrology of the Willamette Basin, Oregon: U.S. Geological Survey Scientific Investigations Report 2005–5168, 83 p.

Cooper, H.H., Jr., 1964, A hypothesis concerning the dynamic balance of fresh water and salt water in a coastal aquifer, *in* Cooper, H.H., Kohout, F.A., Henry, H.R., and Glover, R.E., Sea water in coastal aquifers: U.S. Geological Survey Water-Supply Paper 1613-C, p. 1–12.

Czarnecki, J.B., Clark, B.R., and Reed, T.B., 2003, Conjunctive-use optimization model of the Mississippi River Valley alluvial aquifer of northeastern Arkansas: U.S. Geological Survey Water-Resources Investigations Report 03–4230, 29 p.

Czarnecki, J.B., Clark, B.R., and Stanton, G.P., 2003, Conjunctive-use optimization model of the Mississippi River Valley alluvial aquifer of southeastern Arkansas: U.S. Geological Survey Water-Resources Investigations Report 03–4233, 26 p.

D'Agnese, F.A., O'Brien, G.M., Faunt, C.C., Belcher, W.R., and San Juan, C., 2002, A three-dimensional numerical model of predevelopment conditions in the Death Valley regional ground-water flow system, Nevada and California: U.S. Geological Survey Water-Resources Investigations Report 02–4102, 114 p.

Danskin, W.R., McPherson, K.R., and Woolfenden, L.R., 2006, Hydrology, description of computer models, and evaluation of selected water-management alternatives in the San Bernardino area, California: U.S. Geological Survey Open-File Report 2005–1278, 194 p., 2 pls.

Darton, N.H., 1905, Preliminary report on the geology and underground water resources of the central Great Plains: U.S. Geological Survey Professional Paper 32, 433 p.

Davis, G.H., Green, J.H., Olmsted, F.H., and Brown, D.W., 1959, Ground-water conditions and storage capacity in the San Joaquin Valley, California: U.S. Geological Survey Water-Supply Paper 1469, 287 p., 29 pls.

Dennehy, K.F., 2000, High Plains regional ground-water study: U.S. Geological Survey Fact Sheet 091–00, 6 p.

Dennehy, K.F., 2005, Ground-Water Resources Program: U.S. Geological Survey Fact Sheet 2005–3097, 4 p.

Dennehy, K.F., Litke, D.W., and, McMahon, P.B., 2002, The High Plains aquifer, USA—Groundwater development and sustainability, *in* Hiscock, K.M., Rivett, M.O., and Davison, R.M., eds., Sustainable groundwater development: London, Geological Society, Special Publications, v. 193, p. 99–119.

dePaul, V.T., Rice, D.E., and Zapecza, O.S., 2008, Water-level changes in aquifers of the Atlantic Coastal Plain, predevelopment to 2000: U.S. Geological Survey Scientific Investigations Report 2007–5247, 89 p.

Drost, B.W., Ely, D.M., and Lum, W.E., II, 1999, Conceptual model and numerical simulation of the ground-water-flow system in the unconsolidated sediments of Thurston County, Washington: U.S. Geological Survey Water-Resources Investigations Report 99–4165, 254 p.

Ely, D.M., and Kahle, S.C., 2004, Conceptual model and numerical simulation of the ground-water-flow system in the unconsolidated deposits of the Colville River Watershed, Stevens County, Washington: U.S. Geological Survey Scientific Investigations Report 2004–5237, 72 p.

Feinstein, D., Hart, D., Eaton, T., Krohelski J., and Bradbury, K., 2005, Simulation of regional groundwater flow in southeastern Wisconsin, Report 2—Model results and interpretation: Southeastern Wisconsin Regional Planning Commission, Technical Report 41, 63 p.

Feth, J.H., and others, 1965, Preliminary map of the conterminous United States showing depth to and quality of shallowest ground water containing more than 1,000 parts per million dissolved solids: U.S. Geological Survey Hydrologic Investigations Atlas HA-199, 31 p., 2 pls.

Freckleton, J.R., Martin, Peter, and Nishikawa, Tracy, 1998, Geohydrology of Storage Unit III and a combined flow model of the Santa Barbara and foothill ground-water basins, Santa Barbara County, California: U.S. Geological Survey Water-Resources Investigations Report 97–4121, 80 p.

Galloway, Devin, Jones, D.R., and Ingebritsen, S.E., eds., 1999, Land subsidence in the United States: U.S. Geological Survey Circular 1182, 177 p.

Galloway, Devin, and Riley, F.S., 1999, San Joaquin Valley, California, *in* Galloway, Devin, Jones, D.R., and Ingebritsen, S.E., eds., Land subsidence in the United States: U.S. Geological Survey Circular 1182, p. 23–34.

Gannett, M.W., and Caldwell, R.R., 1998, Geologic framework of the Willamette Lowland aquifer system, Oregon and Washington: U.S. Geological Survey Professional Paper 1424-A, 32 p., 8 pls.

Gannett, M.W., and Lite, K.E., Jr., 2004, Simulation of regional ground-water flow in the upper Deschutes Basin, Oregon: U.S. Geological Survey Water-Resources Investigations Report 03–4195, 84 p.

Gebert, W.A., Graczyk, D.J., and Krug, W.R., 1987, Average annual runoff in the United States, 1951–80: U.S. Geological Survey Hydrologic Investigations Atlas HA-710, 1 sheet, scale 1:7,500,000.

Gebert, W.A., Kennedy, J.L., and Hunt, R.J., 2006, Box 2—The importance of spatial and temporal variability in recharge, *in* Neff, B.P., Piggott, A.R., and Sheets, R.A., 2006, Estimation of shallow ground-water recharge in the Great Lakes Basin: U.S. Geological Survey Scientific Investigations Report 2005–5284, p. 10.

Giese, G.L., Eimers, J.L., and Coble, R.W., 1997, Simulation of ground-water flow in the Coastal Plain aquifer system of North Carolina: U.S. Geological Survey Professional Paper 1404-M, 142 p.

Gilliom, R.J., Barbash, J.E., Crawford, C.G., Hamilton, P.A., Martin, J.D., Nakagaki, Naomi, Nowell, L.H., Scott, J.C., Stackelberg, P.E., Thelin, G.P., and Wolock, D.M., 2006, The quality of our Nation's waters—Pesticides in the Nation's streams and ground water, 1992–2001: U.S. Geological Survey Circular 1291, 172 p.

Gilliom, R.J., Hamilton, P.A., and Miller, T.L., 2001, The National Water-Quality Assessment Program—Entering a new decade of investigations: U.S. Geological Survey Fact Sheet 071–01, 8 p.

Grannemann, N.G., Hunt, R.J., Nicholas, J.R., Reilly, T.E., and Winter, T.C., 2000, The importance of ground water in the Great Lakes region: U.S. Geological Survey Water-Resources Investigations Report 00–4008, 14 p.

Grannemann, N.G., and Reeves, H.W., 2005, Great Lakes Basin water availability and use—A study of the National Assessment of Water Availability and Use Program: U.S. Geological Survey Fact Sheet 2005–3113, 4 p.

Great Valley Center, 1998, Agricultural land conservation in the Great Valley: Great Central Valley agricultural land transaction study, Modesto, CA, 33 p., accessed November 20, 2007, at *http://www.greatvalley.org/publications/agpubs/ag_land_conservation.pdf*

Great Valley Center, 2005, Indicators, The state of the Great Central Valley—The Economy, 1999–2004: State of the Great Central Valley series, Modesto, CA, accessed November 20, 2007, at *http://www.greatvalley.org/indicators*

Ground Water Protection Council, 2007, Ground water report to the Nation—A call to action: Oklahoma City, OK, accessed December 12, 2007, at *http://www.gwpc.org/calltoaction/*

Grubb, H.F., 1998, Summary of hydrology of the regional aquifer systems, Gulf Coastal Plain, South-Central United States: U.S. Geological Survey Professional Paper 1416-A, 61 p.

Gutentag, E.D., Heimes, F.J., Krothe, N.C., Luckey, R.R., and Weeks, J.B., 1984, Geohydrology of the High Plains aquifer in parts of Colorado, Kansas, Nebraska, New Mexico, Oklahoma, South Dakota, Texas, and Wyoming: U.S. Geological Survey Professional Paper 1400-B, 63 p.

Hamilton, P.A., Miller, T.L., and Myers, D.N., 2004, Water quality in the Nation's streams and aquifers—Overview of selected findings, 1991–2001: U.S. Geological Survey Circular 1265, 20 p.

Hanson, R.L., 1991, Evapotranspiration and droughts, *in* National water summary 1988–89—Hydrologic events and floods and droughts: U.S. Geological Water-Supply Paper 2375, p. 99–104.

Hanson, R.T., 2003, Geohydrologic framework of recharge and seawater intrusion in the Pajaro Valley, Santa Cruz and Monterey Counties, California: U.S. Geological Survey Water-Resources Investigations Report 2003–4096, 88 p.

Hanson, R.T., Li, Zhen, and Faunt, C.C., 2004, Documentation of the Santa Clara Valley regional ground-water/surface-water flow model, Santa Clara Valley, California: U.S. Geological Survey Scientific Investigations Report 2004–5231, 85 p.

Hanson, R.T., Martin, Peter, and Koczot, K.M., 2003, Simulation of ground-water/surface-water flow in the Santa Clara-Calleguas ground-water basin, Ventura County, California: U.S. Geological Survey Water-Resources Investigations Report 2002–4136 , 214 p.

Harrill, J.R., and Prudic, D.E., 1998, Aquifer systems in the Great Basin region of Nevada, Utah, and adjacent States—Summary report: U.S. Geological Survey Professional Paper 1409-A, 66 p.

Haugh, C.J., 2002, Hydrogeology and ground-water-flow simulation of the Cave Springs area, Hixson, Tennessee: U.S. Geological Survey Water-Resources Investigations Report 02–4091, 57 p.

Healy, R.W., Winter, T.C., LaBaugh, J.W., and Franke, O.L., 2007, Water budgets—Foundations for effective water-resources and environmental management: U.S. Geological Survey Circular 1308, 90 p.

Heath, R.C., 1983, Ground-water regions of the United States: U.S. Geological Survey Water-Supply Paper 2242, 78 p.

Hodgkins, G.A., Dudley, R.W., and Aichele, S.S., 2007, Historical changes in precipitation and streamflow in the U.S. Great Lakes Basin, 1915–2004: U.S. Geological Survey Scientific Investigations Report 2007–5118, 31 p.

Hsieh, P.A., Barber, M.E., Contor, B.A., Hossain, M.A., Johnson, G.S., Jones, J.L., and Wylie, A.H., 2007, Ground-water flow model for the Spokane Valley–Rathdrum Prairie aquifer, Spokane County, Washington, and Bonner and Kootenai Counties, Idaho: U.S. Geological Survey Scientific Investigations Report 2007–5044, 78 p.

Hutson, S.S., Barber, N.L., Kenny, J.F., Linsey, K.S., Lumia, D.S., and Maupin, M.A., 2004, Estimated use of water in the United States in 2000: U.S. Geological Survey Circular 1268, 46 p.

Idaho Department of Water Resources, 1999, Feasibility of large-scale managed recharge of the Eastern Snake Plain aquifer system: Idaho Department of Water Resources, 244 p.

Izuka, S.K., and Gingerich, S.B., 1998, Ground water in the southern Lihue Basin, Kauai, Hawaii: U.S. Geological Survey Water-Resources Investigations Report 98–4031, 71 p.

Izuka, S.K., and Oki, D.S., 2002, Numerical simulation of ground-water withdrawals in the southern Lihue Basin: U.S. Geological Survey Water-Resources Investigations Report 01–4200, 54 p.

Jennings, C.W., 1977, Geologic map of California: Sacramento, CA, California Division of Mines and Geology, 1:750,000.

Johnson, W.D., 1901, The High Plains and their utilization, 21st annual report of the U.S. Geological Survey, part 4—Hydrography: Washington, D.C., Government Printing Office, p. 601–741.

Johnson, W.D., 1902, The High Plains and their utilization, 22nd annual report of the U.S. Geological Survey, part 4—Hydrography: Washington, D.C., Government Printing Office, p. 631–669.

Johnston, R.H., 1999, Hydrologic budgets of regional aquifer systems of the United States for predevelopment and development conditions: U.S. Geological Survey Professional Paper 1425, 34 p.

Johnston, R.H., and Bush, P.W., 1988, Summary of the hydrology of the Floridan aquifer system in Florida and in parts of Georgia, South Carolina, and Alabama: U.S. Geological Survey Professional Paper 1403-A, 24 p.

Kasmarek, M.C., and Robinson, J.L., 2004, Hydrogeology and simulation of ground-water flow and land-surface subsidence in the northern part of the Gulf Coast aquifer system: U.S. Geological Survey Scientific Investigations Report 2004–5102, 111 p.

Knochenmus, L.A., 2006, Regional evaluation of the hydrogeologic framework, hydraulic properties, and chemical characteristics of the intermediate aquifer system underlying southern west-central Florida: U.S. Geological Survey Scientific Investigations Report 2006–5013, 52 p.

Knowles, N., Dettinger, M., and Cayan, D., 2006, Trends in snowfall versus rainfall for the western United States: Journal of Climate, v. 19, no. 18, p. 4545–4559.

Konrad, C.P., Drost, B.W., and Wagner, R.J., 2005, Hydrogeology of the unconsolidated sediments, water quality, and ground-water/surface-water exchanges in the Methow River Basin, Okanogan County, Washington: U.S. Geological Survey Water-Resources Investigations Report 03–4244, version 1.1, 137 p., 1 pl.

Langevin, C.D., 2001, Simulation of ground-water discharge to Biscayne Bay, southeastern Florida: U.S. Geological Survey Water-Resources Investigations Report 00–4251, 127 p., 3 pls.

Leighton, D.A., and Phillips, S.P., 2003, Simulation of ground-water flow and land subsidence in the Antelope Valley ground-water basin, California: U.S. Geological Survey Water-Resources Investigations Report 2003–4016, 118 p.

Lindgren, R.J., Dutton, A.R., Hovorka, S.D., Worthington, S.R.H., and Painter, Scott, 2004, Conceptualization and simulation of the Edwards aquifer, San Antonio region, Texas: U.S. Geological Survey Scientific Investigations Report 2004–5277, 143 p., 7 pls.

Lindholm, G.F., 1996, Summary of the Snake River Plain Regional Aquifer-System Analysis in Idaho and eastern Oregon: U.S. Geological Survey Professional Paper 1408-A, 59 p.

Lohman, S.W., 1953, High Plains of the West-Central United States, general aspects, ch. 4 of the subsurface facilities of water management and patterns of supply-type area studies, v. 4 of the physical and economic foundation of natural resources: U.S. 83rd Congress, House Committee of Interior and Insular Affairs, p. 70–78.

Lohman, S.W., 1972, Ground-water hydraulics: U.S. Geological Survey Professional Paper 708, 70 p.

Lohman, S.W., 1986, Tribute to Oscar Edward Meinzer, *in* Rosenshein, J.S., Moore, J.E., Lohman, S.W., and Chase, E.B., eds., 1986, Two-hundred years of hydrogeology in the United States: U.S. Geological Survey Open-File Report 86–480, p. 51–60.

Luckey, R.R., and Becker M.F., 1999, Hydrogeology, water use, and simulation of flow in the High Plains aquifer in the northwestern Oklahoma, southeastern Colorado, southwestern Kansas, northeastern New Mexico, and northwestern Texas: U.S. Geological Survey Water-Resources Investigations Report 99–4104, 68 p.

Luckey, R.R., Gutentag, E.D., Heimes, F.J., and Weeks, J.B., 1986, Digital simulation of ground-water flow in the High Plains aquifer in parts of Colorado, Kansas, Nebraska, New Mexico, Oklahoma, South Dakota, Texas, and Wyoming: U.S. Geological Survey Professional Paper 1400-D, 57 p.

Luckey, R.R., Gutentag, E.D., Heimes, F.J., and Weeks, J.B., 1988, Effects of future ground-water pumpage on the High Plains aquifer in parts of Colorado, Kansas, Nebraska, New Mexico, Oklahoma, South Dakota, Texas, and Wyoming: U.S. Geological Survey Professional Paper 1400-E, 44 p.

Martin, A., Jr., and Whiteman, C.D., Jr., 1999, Hydrology of the coastal lowlands aquifer system in parts of Alabama, Florida, Louisiana, and Mississippi: U.S. Geological Survey Professional Paper 1416-H, 51 p., 8 pls.

Maupin, M.A., and Barber, N.L., 2005, Estimated withdrawals from principal aquifers in the United States, 2000: U.S. Geological Survey Circular 1279, 46 p.

McAda, D.P., and Barroll, Peggy, 2002, Simulation of ground-water flow in the Middle Rio Grande Basin between Cochiti and San Acacia, New Mexico: U.S. Geological Survey Water-Resources Investigations Report 02–4200, 81 p.

McGuinness, C.L., 1951, The water situation in the United States with special reference to ground water: U.S. Geological Survey Circular 114, 127 p.

McGuinness, C.L., 1963, The role of ground water in the national water situation: U.S. Geological Survey Water-Supply Paper 1800, 1,121 p.

McGuire, V.L., 2007, Water-level changes in the High Plains aquifer, predevelopment to 2005 and 2003 to 2005: U.S. Geological Survey Scientific Investigations Report 2006–5324, 7 p.

McKee, P.W., and Clark, B.R., 2003, Development and calibration of a ground-water flow model for the Sparta aquifer of southeastern Arkansas and north-central Louisiana and simulated response to withdrawals, 1998–2027: U.S. Geological Survey Water-Resources Investigations Report 03–4132, 71 p., 1 CD–ROM.

McKee, P.W., Clark, B.R., and Czarnecki, J.B., 2004, Conjunctive-use optimization model and sustainable-yield estimation for the Sparta aquifer of southeastern Arkansas and north-central Louisiana: U.S. Geological Survey Water-Resources Investigations Report 03–4231, 30 p.

McMahon, P.B., Dennehy, K.F., Bruce, B.W., Böhlke, J.K., Michel, R.L., Gurdak, J.J., and Hurlbut, D.B., 2006, Storage and transit time of chemicals in thick unsaturated zones under rangeland and irrigated cropland, High Plains, United States: Water Resources Research, v. 42, W03413, doi:10.1029/2005WR004417.

McMahon, P.B., Dennehy, K.F., Bruce, B.W., Gurdak, J.J., and Qi, S.L., 2007, Water-quality assessment of the High Plains aquifer, 1999–2004: U.S. Geological Survey Professional Paper 1749, 136 p.

Meinzer, O.E., 1923, The occurrence of ground water in the United States with a discussion of principles: U.S. Geological Survey Water-Supply Paper 489, 321 p.

Miller, J.A., 1992, Summary of the hydrology of the Southeastern Coastal Plain aquifer system in Mississippi, Alabama, Georgia, and South Carolina: U.S. Geological Survey Professional Paper 1410-A, 38 p.

Miller, J.A., ed., 2000, Ground water atlas of the United States: U.S. Geological Survey Hydrologic Atlas 730, Chapters B–N published separately.

Morgan, D.S., and McFarland, W.D., 1996, Simulation analysis of the ground-water flow system in the Portland Basin, Oregon and Washington: U.S. Geological Survey Water-Supply Paper 2470-B, 83 p.

Mote, P.W., Hamlet, A.F., Clark, M.P., and Lettenmaier, D.P., 2005, Declining mountain snowpack in western North America: Bulletin of the American Meteorological Society, v. 86, no. 1, p. 39–49.

Nace, R.L., 1960, Water management, agriculture, and ground-water supplies: U.S. Geological Survey Circular 415, 12 p.

National Research Council, 2002, Estimating water use in the United States—A new paradigm for the National Water-Use Information Program: Washington, D.C., National Academy Press, 176 p.

National Science and Technology Council, Committee on Environment and Natural Resources, Subcommittee on Water Availability and Quality, 2004, Science and technology to support fresh water availability in the United States: 19 p., accessed December 12, 2007, at *http://www.ostp.gov/nstc/html/swaqreport_2-1-05.pdf*

National Science and Technology Council, Committee on Environment and Natural Resources, Subcommittee on Water Availability and Quality, 2007, A strategy for federal science and technology to support water availability and quality in the United States: 35 p., accessed December 12, 2007, at *http://www.ostp.gov/nstc/html/Fed%20S&T%20Strategy%20for%20Water%209-07%20FINAL.pdf*

Neff, B.P., Day, S.M., Piggott, A.R., and Fuller, L.M., 2005, Base flow in the Great Lakes Basin: U.S. Geological Survey Scientific Investigations Report 2005–5217, 23 p.

Neff, B.P., Piggott, A.R., and Sheets, R.A., 2006, Estimation of shallow ground-water recharge in the Great Lakes Basin: U.S. Geological Survey Scientific Investigations Report 2005–5284, 20 p.

Nichols, W.D., Shade, P.J., and Hunt, C.D., Jr., 1996, Summary of the Oahu, Hawaii, Regional Aquifer-System Analysis: U.S. Geological Survey Professional Paper 1412-A, 61 p.

Nishikawa, Tracy, Izbicki, J.A., Hevesi, J.A., Stamos, C.L., and Martin, Peter, 2005, Evaluation of geohydrologic framework, recharge estimates and ground-water flow of the Joshua Tree area, San Bernardino County, California: U.S. Geological Survey Scientific Investigations Report 2004–5267, 127 p.

Oki, D.S., 1997, Geohydrology and numerical simulation of the ground-water flow system of Molokai, Hawaii: U.S. Geological Survey Water-Resources Investigations Report 97–4176, 62 p.

Oki, D.S., 1999, Geohydrology and numerical simulation of the ground-water flow system of Kona, island of Hawaii: U.S. Geological Survey Water-Resources Investigations Report 99–4073, 70 p.

Oki, D.S., 2002, Reassessment of ground-water recharge and simulated ground-water availability for the Hawi area of north Kohala, Hawaii: U.S. Geological Survey Water-Resources Investigations Report 02–4006, 62 p.

Oki, D.S., 2005, Numerical simulation of the effects of low-permeability valley-fill barriers and the redistribution of ground-water withdrawals in the Pearl Harbor area, Oahu, Hawaii: U.S. Geological Survey Scientific Investigations Report 2005–5253, 111 p.

Oki, D.S., 2006, Numerical simulation of the hydrologic effects of redistributed and additional ground-water withdrawal, Island of Molokai, Hawaii: U.S. Geological Survey Scientific Investigations Report 2006–5177, 49 p.

Olmsted, F.H., and Davis, G.H., 1961, Geologic features and ground-water storage capacity of the Sacramento Valley, California: U.S. Geological Survey Water-Supply Paper 1497, 241 p., 5 pls.

Page, R.W., 1973, Base of fresh ground water (approximately 3,000 micromhos) in the San Joaquin Valley, California: U.S. Geological Survey Hydrologic Investigations Atlas HA-489, 1 sheet.

Page, R.W., 1986, Geology of the fresh ground-water basin, Central Valley, California, with texture maps and sections: U.S. Geological Survey Professional Paper 1401-C, 54 p., 5 pls.

Payne, D.F., Provost, A.M., Painter, J.A., Rumman, M.A., and Cherry, G.S., 2006, Application of ground-water flow and solute-transport models to simulate selected ground-water management scenarios in coastal Georgia and adjacent parts of South Carolina and Florida, 2000–2100: U.S. Geological Survey Scientific Investigations Report 2006–5077, 78 p.

Payne, D.F., Rumman, M.A., and Clarke, J.S., 2005, Simulation of ground-water flow in coastal Georgia and adjacent parts of South Carolina and Florida—Predevelopment, 1980, and 2000: U.S. Geological Survey Scientific Investigations Report 2005–5089, 91 p.

Plummer, L.N., Bexfield, L.M., Anderholm, S.K., Sanford, W.E., and Busenberg, Eurybiades, 2004, Geochemical characterization of ground-water flow in the Santa Fe Group aquifer system, Middle Rio Grande Basin, New Mexico: U.S. Geological Survey Water-Resources Investigations Report 03–4131, 395 p.

Pool, D.R., and Dickinson, J.E., 2007, Ground-water flow model of the Sierra Vista Subwatershed and Sonoran portions of the Upper San Pedro Basin, southeastern Arizona, United States, and northern Sonora, Mexico: U.S. Geological Survey Scientific Investigations Report 2006–5228, 48 p.

Pool, D.R., and Eychaner, J.H., 1995, Measurements of aquifer-storage change and specific yield using gravity surveys: Ground Water, v. 33, no. 3, p. 425–432.

Provost, A.M., Payne, D.F., and Voss, C.I., 2006, Simulation of saltwater movement in the Upper Floridan aquifer in the Savannah, Georgia–Hilton Head Island, South Carolina, area, predevelopment–2004, and projected movement for 2000 pumping conditions: U.S. Geological Survey Scientific Investigations Report 2006–5058, 124 p.

Prudic, D.E., Konikow, L.F., and Banta, E.A., 2004, A new Streamflow-Routing (SFR1) Package to simulate stream-aquifer interaction with MODFLOW-2000: U.S. Geological Survey Open-File Report 04–1042, 95 p.

Qi, S.L., Konduris, Alexandria, Litke, D.W., and Dupree, Jean, 2002, Classification of irrigated land using satellite imagery, the High Plains aquifer, nominal date 1992: U.S. Geological Survey Water-Resources Investigations Report 02–4236, 31 p.

Randall, A.D., 2001, Hydrogeologic framework of stratified-drift aquifers in the glaciated Northeastern United States: U.S. Geological Survey Professional Paper 1415-B, 179 p., 1 pl.

Reed, T.B., 2003, Recalibration of a ground-water flow model of the Mississippi River Valley alluvial aquifer of northeastern Arkansas, 1918–1998, with simulations of water levels caused by projected ground-water withdrawals through 2049: U.S. Geological Survey Water-Resources Investigations Report 03–4109, 58 p.

Reeves, H.W., Wright, K.V., and Nicholas, J.R., 2004, Hydrogeology and simulation of regional ground-water-level declines in Monroe County, Michigan: U.S. Geological Survey Water-Resources Investigations Report 2003–4312, 72 p.

Reichard, E.G., Land, Michael, Crawford, S.M., Johnson, Tyler, Everett, R.R., Kulshan, T.V., Ponti, D.J., Halford, K.J., Johnson, T.A., Paybins, K.S., and Nishikawa, Tracy, 2003, Geohydrology, geochemistry, and ground-water simulation-optimization of the Central and West Coast Basins, Los Angeles County, California: U.S. Geological Survey Water-Resources Investigations Report 2003–4065, 184 p.

Roy, S.B., Ricci, P.F., Summers, K.V., Chung, C.F., and Goldstein, R.A., 2005, Evaluation of the sustainability of water withdrawals in the United States, 1995 to 2025: Journal of the American Water Resources Association, v. 41, no. 5, p. 1091–1108.

Ryker, S.J., 2001, Mapping arsenic in groundwater: Geotimes v. 46, no. 11, p. 34–36; updated map accessed July 17, 2006, at *http://water.usgs.gov/nawqa/trace/pubs/geo_v46n11/fig1.html*

Scanlon, B.R., Reedy, R.C., Stonestrom, D.A., Prudic, D.E., and Dennehy, K.F., 2005, Impact of land use and land cover change on groundwater recharge and quality in the Southwestern U.S.: Global Change Biology, v. 11, no. 10, p. 1577–1593.

Schmid, Wolfgang, Hanson, R.T., Maddock, Thomas, III, and Leake, S.A., 2006, User guide for the Farm Process (FMP1) for the U.S. Geological Survey's modular three-dimensional finite-difference ground-water flow model, MODFLOW-2000: U.S. Geological Survey Techniques and Methods 6-A17, 127 p.

Sepulveda, Nicasio, 2002, Simulation of ground-water flow in the intermediate and Floridan aquifer systems in peninsular Florida: U.S. Geological Survey Water-Resources Investigations Report 02–4009, 130 p.

Shedlock, R.J., and Bolton, D.W., 2006, Sustainability of the ground-water resources in the Atlantic Coastal Plain of Maryland: U.S. Geological Survey Fact Sheet 2006–3009, 2 p.

Sheets, R.A., and Simonson, L.A., 2006, Compilation of regional ground-water divides for principal aquifers corresponding to the Great Lakes Basin, United States: U.S. Geological Survey Scientific Investigations Report 2006–5102, 23 p.

Solley, W.B., Pierce, R.R., and Perlman, H.A., 1998, Estimated use of water in the United States in 1995: U.S. Geological Survey Circular 1200, 71 p.

Stamos, C.L., Martin, Peter, Nishikawa, Tracy, and Cox, B.F., 2001, Simulation of ground-water flow in the Mojave River Basin, California: U.S. Geological Survey Water-Resources Investigations Report 2001–4002, 129 p.

Stanton, G.P., and Clark, B.R., 2003, Recalibration of a ground-water flow model of the Mississippi River Valley alluvial aquifer in southeastern Arkansas, 1918–1998, with simulations of hydraulic heads caused by projected ground-water withdrawals through 2049: U.S. Geological Survey Water-Resources Investigations Report 03–4232, 48 p.

Sun, R.J., and Johnston, R.H., 1994, Regional Aquifer-System Analysis Program of the U.S. Geological Survey, 1978–1992: U.S. Geological Survey Circular 1099, 126 p.

Swain, L.A., Mesko, T.O., and Hollyday, E.F., 2004, Summary of the hydrogeology of the Valley and Ridge, Blue Ridge, and Piedmont physiographic provinces in the Eastern United States: U.S. Geological Survey Professional Paper 1422-A, 23 p.

Swenson, S., Wahr, J., and Milly, P.C.D., 2003, Estimated accuracies of regional water storage variations inferred from the Gravity Recovery and Climate Experiment (Grace): Water Resources Research, v. 39, no. 8, 1223, doi:10.1029/2002WR001808.

Taylor, C.J., and Alley, W.M., 2001, Ground-water-level monitoring and the importance of long-term water-level data: U.S. Geological Survey Circular 1217, 68 p.

The H. John Heinz III Center for Science, Economics, and the Environment, 2002, The state of the Nation's ecosystems, measuring the lands, waters, and living resources of the United States: Cambridge, UK, Cambridge University Press, 270 p.

The H. John Heinz III Center for Science, Economics, and the Environment, 2006, Filling the gaps—Priority data needs and key management challenges for national reporting on ecosystem condition: The H. John Heinz III Center for Science, Economics and the Environment, Washington D.C., 104 p.

Theis, C.V., 1940, The source of water derived from wells: Civil Engineering, v. 10, no. 5, p. 277–280.

Thiros, S.A., 2006, Evaluation of the ground-water flow model for Northern Utah Valley, Utah, updated to conditions through 2002: U.S. Geological Survey Scientific Investigations Report 2006–5064, 20 p.

Thomas, H.E., 1951, The conservation of ground water—A survey of the present ground-water situation in the United States: New York, McGraw-Hill, 321 p.

Thornthwaite, C.W., 1948, An approach toward a rational classification of climate: Geographical Review, v. 38, no. 1, p. 55–94.

Trapp, Henry, Jr., and Horn, M.A., 1997, Ground water atlas of the United States—Delaware, Maryland, New Jersey, North Carolina, Pennsylvania, Virginia, West Virginia: U.S. Geological Survey Hydrologic Atlas, HA 730-L, 24 p.

Trapp, Henry, Jr., and Meisler, Harold, 1992, The regional aquifer system underlying the Northern Atlantic Coastal Plain in parts of North Carolina, Virginia, Maryland, Delaware, New Jersey, and New York—Summary: U.S. Geological Survey Professional Paper 1404-A, 33 p., 11 pls.

Tucci, Patrick, and McKay, R.M., 2006, Hydrogeology and simulation of ground-water flow in the Silurian–Devonian aquifer system, Johnson County, Iowa: U.S. Geological Survey Scientific Investigations Report 2005–5266, 73 p.

Underwood, M.R., Meyer, William, and Souza, W.R., 1995, Ground-water availability from the Hawi aquifer in the Kohala area, Hawaii: U.S. Geological Survey Water-Resources Investigations Report 95–4113, 57 p.

U.S. Department of Agriculture, 1999, 1997 census of agriculture, geographic area series: National Agricultural Statistics Service, accessed November 20, 2007, at *http://www.census.gov/prod/www/abs/agr-geo.html*

U.S. Department of Commerce, 1968, Climate atlas of the United States: Washington D.C., Environmental Science Services Administration, Environmental Data Services, 80 p.

U.S. Geological Survey, 1984, National water summary 1983—Hydrologic events and issues: U.S. Geological Survey Water-Supply Paper 2250, 243 p.

U.S. Geological Survey, 1985, National water summary 1984—Hydrologic events, selected water-quality trends, and ground-water resources: U.S. Geological Survey Water-Supply Paper 2275, 467 p.

U.S. Geological Survey, 1998, Strategic directions for the U.S. Geological Survey Ground-Water Resources Program—A report to Congress: U.S. Geological Survey, 14 p.

U.S. Geological Survey, 2003, Principal aquifers of the 48 conterminous United States, Hawaii, Puerto Rico, and the U.S. Virgin Islands: U.S. Geological Survey, accessed November 20, 2007, at *http://www.nationalatlas.gov/mld/aquifrp.html*

U.S. Geological Survey, 2007, Facing tomorrow's challenges—U.S. Geological Survey science in the decade 2007–2017: U.S. Geological Survey Circular 1309, 70 p.

U.S. Water Resources Council, 1978, The Nation's water resources, 1975–2000, v. 1—Summary: Washington, D.C., U.S. Government Printing Office.

U.S. Water Resources Council, 1980, Essentials of ground-water hydrology pertinent to water-resources planning: Washington, D.C., U.S. Government Printing Office, U.S. Water Resources Council Hydrology Committee Bulletin 16 (revised), 38 p.

Vaccaro, J.J., 1999, Summary of the Columbia Plateau Regional Aquifer-System Analysis, Washington, Oregon, and Idaho: U.S. Geological Survey Professional Paper 1413-A, 51 p.

Vaccaro, J.J., Hansen, A.J., Jr., and Jones, M.A., 1998, Hydrogeologic framework of the Puget Sound aquifer system, Washington and British Columbia: U.S. Geological Survey Professional Paper 1424-D, 77 p.

Voronin, L.M., 2004, Documentation of revisions to the Regional Aquifer System Analysis model of the New Jersey Coastal Plain: U.S. Geological Survey Water-Resources Investigations Report 03–4268, 49 p., 1 pl., 1 CD–ROM.

Warner, K.L., and Arnold, T.L., 2005, Framework for regional synthesis of water-quality data for the glacial aquifer system in the United States: U.S. Geological Survey Scientific Investigations Report 2005–5223, 6 p.

Weeks, J.B., Gutentag, E.D., Heimes, F.J., and Luckey, R.R., 1988, Summary of the High Plains Regional Aquifer-System Analysis in parts of Colorado, Kansas, Nebraska, New Mexico, Oklahoma, South Dakota, Texas, and Wyoming: U.S. Geological Survey Professional Paper 1400-A, 30 p.

Weissmann, G.S., Bennett, G.L., and Lansdale, A.L., 2005, Factors controlling sequence development on Quaternary fluvial fans, San Joaquin Basin, California, USA, *in* Harvey, A.M., Mather, A.E., and Stokes, M., eds., Alluvial fans—Geomorphology, sedimentology, dynamics: The Geological Society of London, Special Publications 251, p. 169–186.

Westjohn, D.B., and Weaver, T.L., 1998, Hydrogeologic framework of the Michigan Basin regional aquifer system: U.S. Geological Survey Professional Paper 1418, 45 p.

Wilkins, D.W., 1998, Summary of the Southwest alluvial basins Regional Aquifer-System Analysis in parts of Colorado, New Mexico, and Texas: U.S. Geological Survey Professional Paper 1407-A, 49 p.

Williamson, A.K., Prudic, D.E., and Swain, L.A., 1989, Ground-water flow in the Central Valley, California: U.S. Geological Survey Professional Paper 1401-D, 127 p.

Wilson, W.E., and Moore, J.E., eds., 1998, Glossary of hydrology: Alexandria, VA, American Geological Institute, 248 p.

Wolock, D.M., 2003a, Estimated mean annual natural ground-water recharge in the conterminous United States: U.S. Geological Survey Open-File Report 03–311, digital dataset, accessed on November 20, 2007, at *http://water.usgs.gov/lookup/getspatial?rech48grd*

Wolock, D.M., 2003b, Base-flow index grid for the conterminous United States: U.S. Geological Survey Open-File Report 03–263, digital dataset, accessed on November 20, 2007, at *http://water.usgs.gov/lookup/getspatial?bfi48grd*

Woodward, D.G., and Anderson, H.W., Jr., 1986, Hydrogeologic and water-quality characteristics of the Cretaceous aquifer, southwestern Minnesota: U.S. Geological Survey Water-Resources Investigations Report 84–4153, 2 sheets.

Young, H.L., 1992, Summary of ground-water hydrology of the Cambrian–Ordovician aquifer system in the northern Midwest, United States: U.S. Geological Survey Professional Paper 1405-A, 55 p.

Zogorski, J.S., Carter, J.M., Ivahnenko, Tamara, Lapham, W.W., Moran, M.J., Rowe, B.L., Squillace, P.J., and Toccalino, P.L., 2006, The quality of our Nation's waters—Volatile organic compounds in the Nation's ground water and drinking-water supply wells: U.S. Geological Survey Circular 1292, 101 p.